REWIRING
DEMOCRACY

Strong Ideas Series
Edited by David Weinberger

The Strong Ideas Series explores the latest ideas about how technology is affecting culture, business, science, and everyday life. Written for general readers by leading technology thinkers and makers, books in this series advance provocative hypotheses about the meaning of new technologies for contemporary society.

The Strong Ideas Series is published with the generous support of the MIT Libraries.

Hacking Life: Systematized Living and Its Discontents, Joseph M. Reagle, Jr.

The Smart Enough City: Putting Technology in Its Place to Reclaim Our Urban Future, Ben Green

Sharenthood: Why We Should Think Before We Post About Our Kids, Leah A. Plunkett

Data Feminism, Catherine D'Ignazio and Lauren Klein

Artificial Communication: How Algorithms Produce Social Intelligence, Elena Esposito

The Digital Closet: How the Internet Became Straight, Alexander Monea

On the Brink of Utopia: Reinventing Innovation to Solve the World's Largest Problems, Thomas Ramge and Rafael Laguna de la Vera

Fantasies of Virtual Reality: Untangling Fiction, Fact, and Threat, Marcus Carter and Ben Egliston

Rewiring Democracy: How AI Will Transform Our Politics, Government, and Citizenship, Bruce Schneier and Nathan E. Sanders

REWIRING DEMOCRACY

HOW AI WILL TRANSFORM OUR POLITICS, GOVERNMENT, AND CITIZENSHIP

BRUCE SCHNEIER AND NATHAN E. SANDERS

THE MIT PRESS   CAMBRIDGE, MASSACHUSETTS   LONDON, ENGLAND

The MIT Press
Massachusetts Institute of Technology
77 Massachusetts Avenue
Cambridge, MA 02139
mitpress.mit.edu

The MIT Press would like to thank the anonymous peer reviewers who provided comments on drafts of this book. The generous work of academic experts is essential for establishing the authority and quality of our publications. We acknowledge with gratitude the contributions of these otherwise uncredited readers.

This book was set in ITC Stone and Avenir by New Best-set Typesetters Ltd. Printed and bound in the United States of America.

Library of Congress Cataloging-in-Publication Data

Names: Schneier, Bruce, 1963– author | Sanders, Nathan E. author
Title: Rewiring democracy : how AI will transform our politics,
   government, and citizenship / Bruce Schneier and
   Nathan E. Sanders.
Description: Cambridge, Massachusetts : The MIT Press, [2025] |
   Series: Strong ideas | Includes bibliographical references and index.
Identifiers: LCCN 2025023676 (print) | LCCN 2025023677 (ebook) |
   ISBN 9780262049948 hardcover | ISBN 9780262384407 epub |
   ISBN 9780262384414 pdf
Subjects: LCSH: Democracy—Technological innovations |
   Artificial intelligence—Political aspects
Classification: LCC JC423 .S356 2025 (print) | LCC JC423 (ebook)
LC record available at https://lccn.loc.gov/2025023676
LC ebook record available at https://lccn.loc.gov/2025023677

10 9 8 7 6 5 4 3 2 1

EU Authorised Representative: Easy Access System Europe, Mustamäe tee 50, 10621 Tallinn, Estonia | Email: gpsr.requests@easproject.com

To Martin Schneier and to Neill Fred Sanders

CONTENTS

PREFACE

When we started writing this book in early 2024, the big questions about artificial intelligence and democracy felt largely hypothetical. Could propagandists use AI deepfakes to swing an election? Could police use AI facial recognition for mass surveillance? Could AI be used to perform the tasks of civil servants? We found that, even then, these questions were not really speculative.

People and governments around the world of all political stripes have been adopting AI across all these use cases and more, as demonstrated by the real-world examples we'll discuss in this book, and this trend has accelerated. It's now clear that AI *could* impact wide-ranging aspects of democracy. But what remains are three basic questions: *Should* we adopt AI in these areas? What range of effects should we allow it to have? How can we make it work to the benefit of democracy?

We're finishing this book at a time when global democracy is being roiled by the words and actions of the second Trump administration. US institutions are being handed over to a tech plutocrat, Elon Musk, who is carrying out an explicit plan to replace humans in government with an "AI-first strategy."[1]

Vice President J.D. Vance just returned from Europe, where he admonished the EU to eliminate its newly adopted AI regulations, saying, "The AI future is not going to be won by hand-wringing about safety."[2] And the biggest AI developers are aggressively lobbying the new administration to not regulate them. OpenAI recently framed the issue in urgent terms of competition against China, urging the US to gain advantage through "an ambitious government adoption strategy."[3]

We can't know what will develop by the time you read this, and this book deliberately takes a more international and longer-term view than these events. But watching the impact this technology is having right now has made it clear that AI will play a central role in the next stage of democratic transformation in the US and across the world. Our goal with this book is to influence the public conception of AI and democracy to help steer that transformation away from disaster and authoritarianism and towards progress and pluralism.

Cambridge, MA, April 2025

INTRODUCTION

1

ARTIFICIAL INTELLIGENCE MEETS DEMOCRACY

Once there was a group of ardent and committed citizens who did not trust each other to administer the systems of their own democracy. They feared that lawmakers could be corrupted, judges could be bribed, and voters could be dazzled by demagogues.

Their solution was to entrust some of the most critical decisions of governing to a machine instead of to people. They built intricate technologies, using mathematical principles, for choosing who among them would serve in public magistracies and sit on juries. They established a sort of ID card that was distributed to all eligible citizens and served as a ticket for inclusion in their automated selection process. When concerns arose that the machines could be rigged, they engineered a way to make them more hack-proof by doubling the number of machines and running them in parallel.

This is not a modern-day story. It comes from Athens during the time of Aristotle, circa the fourth century BCE. The selection machine was called a *kleroterion*. It was a large stone stele encased in a wooden frame enclosing a system of tubes and dice. Their principle of fairness was randomization: the *kleroterion* was an early lottery drawing device, not too different from the ping-pong ball–style mechanical mixers used in the Eurojackpot or American Powerball.[1] The tickets were called *pinakia* and were typically made of bronze. The system of using two *kleroteria* in tandem was called *synclerosis*, or simultaneous allotment.[2] The Athenians used this technology to give everyone (well, every free adult male citizen) an equal chance at decision-making.

Democracy has always been intertwined with technology. Machines like the *kleroterion* and the modern voting booth help democracies function. General-purpose technologies, like railroads, television, and the internet, have had pervasive impacts on democracies. As the political theorist David Runciman says, the state itself is a sort of machine: an artificial agent for governing our lives and actions. So are corporations.[3]

Today, the technologies affecting democracy are more digital, more connected, and more complex than ever before. In recent decades, nations worldwide have grappled with the effects of technologies like social media and drone warfare. At the same time, innovators like Estonia and Taiwan have created digital identity schemes[4] and innovative uses of web platforms for deliberative assemblies.[5] As surely as technology will keep evolving, so will democracy.

Meanwhile, democracy is facing real challenges worldwide: eroding democratic institutions, rising authoritarianism, worsening climate change, surging wealth inequalities, social dislocation, and more. Technology is a contributor to many of these problems, and while it may not solve any of them, it will definitely impact how they play out.

Artificial intelligence is poised to change democracy yet again. AI-generated misinformation spreading on social media, deepfake propaganda disrupting political campaigns, and AI being applied in government and making mistakes that cause harm are all real concerns, but just a small part of a much richer story.

The impacts of AI on democracy will be far larger than those of other recent technological transformations, such as social media. AI has a pluripotent quality that is more akin to money's role in politics and governance, because it has the power to turn speech into action. AI agents can take action autonomously after receiving initial instructions from a person, or from another AI. Anyone capable of speaking (or typing) a command can compel an AI agent, or a thousand of them, to make it so. Today, the scope of those actions is limited; AI can send an email, order a pizza, or trigger your smart thermostat to turn on. In the future, as we connect more and more devices and processes to an internet that includes AI agents, that scope will grow. If laws permit it, AI will be capable of lobbying a legislator, filing a lawsuit, or casting a vote.

As technologists who have one foot in academia and the other in industry, we see numerous ways that AI is already starting to infiltrate, influence, and impact all the activities of democracy: political campaigning, legislation, government administration, arbitration, advocacy, conducting elections, and more. Many are nuanced uses that don't evoke obvious fear or excitement, even if their implications are potentially profound. Taken together, they constitute a new future for democracy, a very different one from what we would have imagined even a few years ago.

One of our first encounters with the strange new world of AI-infused democracy came in January 2023, just weeks after the public release of ChatGPT. We had written an article in the *New York Times* about the potential for this new tool to supercharge lobbying.[6] In putting a spotlight on how elites might leverage AI to generate new forms of influence over public policy, we anticipated rebukes from lobbyists. Unsettlingly, the most prominent response came from ChatGPT itself. A few days later, the *Times* published a letter generated by the AI—a first for the paper—rebutting the concerns outlined in our article.[7] We were suddenly no longer engaged in political debate about AI, but also with AI.

All our experiences of the twenty-first century will be, in no small part, a battle over how to rewire democracy using AI. Entrenched elected officials, political movements with authoritarian tendencies, and the billionaire class all regard AI as a new tool to consolidate and centralize power. But the rest of us, the public, can harness it as a tool to distribute power instead.

This book seeks to anticipate what people on all sides will do to introduce AI into democratic processes in the coming years and to illuminate how best to steer the future of democracy towards participatory, pluralistic, and truly democratic ends.

We recognize the generational challenges facing democracies in the modern world, so we see the potential for AI to hasten the erosion of democratic systems and values. Nonetheless, we're optimists about democracy as an evolving, self-correcting form of government and are confident of finding ways to leverage new tools to make it work better. We're interested in opportunities for AI to be used in ways that promote democracy, and in calling out what needs to change to make those benefits real.

Our perspectives on how AI technologies are being developed and used in governance today have been formed from diverse experiences. Bruce Schneier is a cybersecurity expert, with decades spent thinking about socio-technical systems and how governments should use them responsibly. He views democracy as an information system, and has annually convened groups of experts to discuss how to reimagine democracy from the ground up.[8] The discussions and outputs of those meetings have been fascinating, and informed much of this book.

Nathan E. Sanders is a data scientist with years of experience in putting AI and machine learning tools to work and analyzing their capabilities and limitations. He's built practical systems to address scientific, policy, and business problems across broad domains: environmental justice, public health, astrophysics, online advertising, film production, and drug discovery. He cocreated an online platform to help organize legislative action,[9] used statistical models to brief legislators, and wrote a bill on regulatory data transparency that passed into Massachusetts law.[10]

Together, we've written dozens of articles—for academics and for the public—exploring a myriad of ways that AI is currently impacting and might in the future impact democratic processes. But we have gaps in our experience. We're both white guys living in Cambridge, Massachusetts, conducting research at Harvard University. We've tried to survey the present and future impacts of AI on democracies around the world, but we know our experience is rooted in the US. And while we are engaged citizens, we have not run political campaigns, held elective office, or had legal training.

We're fortunate to be helped by others who have helped us fill in our own gaps. They've engaged in conversations and read drafts of this book to help us see further and understand better. We hope you'll find the products of those conversations as fruitful as we have.

2

HOW AI COULD AFFECT DEMOCRACY

Political actors around the world recognize the transformative potential of AI—unlocked by advancements in computing power, data, and modeling techniques—but they have different ideas about which directions things should go. Politicians and campaigns hope to gain advantages in advertising, fundraising, or messaging by being the first to adopt AI assistive tools. Civic technologists worldwide are trying to use AI to make government more equitable, more efficient, and more responsive. Researchers are building AI-based crowdsourcing tools that solicit policy consensus from human participants online.

Some think that AI itself can suggest new directions for democracy. Countries including Denmark, Japan, the US, and the UK have already seen AI avatars standing for election or forming political parties. Meanwhile, leaders of the military-industrial complex, policy hawks, and Big Tech boosters see AI as a martial tool, critical to winning a post–Cold War arms race between the West and China.

Many of these same tools can be exploited by people who want to make democracy more authoritarian. AI surveillance tools intended to make policing fairer and more accountable can easily be redirected to enforce repression and injustice. AI tools designed to make civil servants more effective can be turned to removing human judgment and compassion from bureaucratic systems. AI is not a solution to the problem of democracies devolving to authoritarianism, but will inhabit the territory between these forces for years to come. To understand the potential impacts of AI on democracy, we must always think beyond the capabilities and innate properties of AI, and focus on the systems, incentives, and political forces within which the AI is built, deployed, and wielded over time.

Most governments are understandably concerned about AI's risks. AI systems can be biased, make mistakes, and be used to facilitate illicit activities. In practically every election of the past few years, there have been urgent concerns over AI deepfakes: synthetically produced images, audio, or video that create the perception a candidate did something they didn't really do, or said something they didn't really say. Some governments have taken concern over AI to the extreme. In 2023, the UK hosted the first global summit on the existential risks—that is, to the very survival of humanity—of AI. Yet these governments' constituents have more immediate fears triggered by technology, like its near-term harms and misuse, and are rightfully skeptical of the motivations of companies and politicians worrying about far-future existential AI risks.

There is another framing of the modern story of AI, one that puts the systemic risks of inequitable outcomes due to bias front and center. This is, in effect, a fight for justice. Computer scientist Joy Buolamwini founded the Algorithmic Justice League to advocate for equitable AI, as does sociologist

Ruha Benjamin's Ida B. Wells Just Data Lab. The movement to make AI more just is large and diverse, led by trailblazers from varied fields, including data journalist Meredith Broussard, sociologist Safiya Umoja Noble, data scientist Rumman Chowdhury, and mathematician Cathy O'Neil. They are taking the right approach by holding corporate power accountable for delivering ethical, fair, and just AI systems.

Meanwhile, governments are only just beginning to regulate the Big Tech companies developing and marketing AI. In 2024, the European Union passed the first comprehensive regulation of AI technology: the EU AI Act. The EU should be lauded for acting where other governments have not, but criticized for its weak protections for human rights and ample loopholes for companies to do as they please.[1] And democratic governments are just beginning to recognize their potential for actively shaping the AI ecosystem. A largely unconstrained industrial sector, supercharged by massive US private capital investment, has raced to a dominant position controlling the most advanced AI models. Echoing the development of the internet, this industry benefits enormously from government-funded basic research, and yet now returns its profits and delegates its values almost exclusively to private entities.[2] So far, only a few governments have extended beyond basic research, seed funding, and computational infrastructure to directly build and provision AI models—that is, to create Public AI.[3]

When new technologies are introduced in governance, they frequently have the effect of concentrating power, and AI will do this efficiently. Consider the technology of government bureaucracies. The size of large agencies endows elected leaders with enormous potency to implement policy, but also necessarily dilutes decision-making across many layers and individuals. Those humans each act from their own remit, perspectives, and (one hopes) ethics. AI offers leaders the centralizing capacity to control government action at an even larger scale, unconstrained by the costs of public servant salaries, with instantaneous coordination and unquestioning compliance. Meanwhile, advancement of AI has emboldened corporate profiteers who have already extracted fortunes from government technology investment, and see orders of magnitude greater opportunities to sell tech products with AI. The implications for leaders with autocratic and fascist interests are profound.

We are still at the start of a long journey in democracies' interaction with AI. The topics that have dominated the public discourse are only a few of the ways AI will impact democracy. AI deepfakes are just the latest version of an age-old political practice. We have been photoshopping and, before that, airbrushing and, before that, staging political images and propaganda for decades. Stalin airbrushed his enemies out of photographs in the 1950s[4] and the US Civil War photographer Alexander Gardner staged powerful battle scenes a century earlier.[5] The technology has changed—and AI image generators are much easier to use than Photoshop—but you have never been completely able to take for granted the images in front of your eyes.

The more interesting changes to democracy from AI will come in the places where few are looking. The early pioneers of radio were not setting out to change how politicians communicate with their electorate, yet politicians gradually found transformative ways to use the new technology. The changes caused by technological advancement tend to come from the bottom up, not the top down. They tend to be incremental; their most significant effects compound over time. This is especially true in an era where technological advancement comes from universities and industry. This is not the era of the Manhattan Project or the Apollo program; an AI arms race metaphor just doesn't make sense.[6]

And yet, technologies tend to push society in particular directions—even when they weren't developed to do so—because of the conditions within which they are developed and used. In the nineteenth century, railroads had enormous potential to connect the disconnected and equalize access to people, places, and power. But their most visible impact was the creation of unprecedented wealth among a new class of oligarchs. In the following century, both the promise and actual impacts of the internet were much the same: many benefited somewhat from connecting through the internet, and a few profited staggeringly from the rest of us using it. These technologies betrayed their promises to the public interest because their value was captured and commandeered by private interests. Today, AI has similar potential to empower, but faces the same risks of subversion to benefit the few instead of the many. The potential capture of AI technologies by a few powerful companies will exacerbate its risks to democracy.

3

WHY AI IS BEING USED

You've probably heard thousands of exciting claims about the tasks modern AI tools can perform. They can write a great term paper. They can plan your vacation. They can even drive cars. Most of those claims are inflected with sales hype, but also contain some truth.

AI is a basket of technologies. No one, especially experts, quite agrees on what does or does not count as AI. For our purposes, we will define AI very generally, as a computer system that is capable of cognitive tasks that have traditionally been within the exclusive purview of humans. It's a broad definition that includes a wide range of different methods, including the rules-based software systems that dominated the AI field of the 1970s, statistical models and probabilistic methods that powered IBM's *Jeopardy!*-playing computer Watson fifteen years ago, the deep neural networks that have dominated the field ever since then, the large language model generative AI systems that burst into public consciousness in 2022 through tools like ChatGPT, and whichever next-generation models are used to perform cognitive tasks.

In this book, we will explore many applications of AI that are still in development or that we see on the horizon, and discuss their potential effects. We will largely assume these applications are indeed possible, and will discuss their likely risks and shortcomings. Today, many AI tools can't accomplish what their creators say they will, and that will probably always be true, but the sphere of AI capabilities is expanding quickly. Even during the year we spent writing this book, we repeatedly replaced future predictions with newly current examples: AIs participating in livestreamed political debates, legislatures building AI models to help draft policy, judges querying AI to help interpret law. And we know that more will happen by the time you read this.

The specifics of which AI tools work well and which are half-baked will change over time. AI will be used to enhance the abilities of and substitute for humans throughout every democratic institution with material effects, whether they work well or not. Some of those effects will be obvious and straightforward, and others will unfold over time.

To understand how and why AI will impact democracy, it is not necessary to understand how these systems work in detail, and we will largely eschew technical jargon. But it is important to understand AI's basic capabilities, because it is from these that applications emerge.

At its core, most AI technologies are predictive. Does this X-ray show a benign or a malignant tumor? Will you arrive at your destination sooner if you turn left or right? Are you more likely to avoid an accident if you slow down or swerve? Generative AI—systems like chatbots or image generators that can produce novel results based on training—is fundamentally predictive, too: Such systems can predict the likely next word in a sentence, or what color a pixel should be in an image. Combinations of these predictive and generative capabilities can be used to accomplish tasks that involve deciding, summarizing, and executing. Decision-making has three parts: enumerating options, predicting their outcomes, and judging their relative merits. AI's ability to process large amounts of information quickly makes it helpful to describe, summarize, and synthesize long documents and large datasets. Depending on what peripherals the AI is hooked up to, it can act: execute a command, send a message, trigger another system. The ability to decide and act means that an AI can operate autonomously, without any humans in the loop. This is the promise of an AI agent: AIs can receive, interpret, and act upon signals from the real world. Add physical capabilities to this, and you have a robot.[1]

AI performs these familiar human tasks differently than people, and differently does not necessarily mean better. AIs can have flawed performance or inherent bias, and can make things up. Humans make mistakes too, of course, but AI mistakes can be very different from the kinds of human mistakes we know how to build systems to mitigate. It can also lead to job loss, as employers choose cheaper AI over the more expensive humans.

In general, AI-generated output doesn't need to be better than what a human would produce to be useful. When considering the potential impacts of AI capabilities, it's important to compare them to their alternatives, not to perfection. The status quo and alternatives to AI automation are often human-mediated systems. The utility of an image-generation AI to a political campaign does not depend on it producing museum-quality art, or works with groundbreaking originality. If it can produce graphics that are good enough for an advertisement, that's sufficient. Similarly, AI can be useful to legislators even if it does not make their legislation better, but merely saves them time drafting it.

You might trust most humans to perform a task more than you would an AI, but that isn't necessarily a useful comparison. A better comparison is between an AI and a human system like a bureaucracy, in which individual people are constrained by a baroque apparatus. Organizational bureaucracies are a familiar way that humans automate tasks, by incorporating multiple people into a single system to implement a central policy across a larger jurisdiction. Whether governmental, corporate, or even religious, bureaucracies have the superhuman quality of effectuating action simultaneously across multiple places and times. And, unlike individual humans, they can last for centuries. But, as we know all too well, bureaucracies are not perfect—just like AIs. It's valid to point out that an AI deciding whether an applicant is eligible for unemployment insurance benefits might make a mistake because it lacks context and understanding that would be obvious to any person. But recognize that humans within bureaucracies often make the same apparently careless, seemingly unaccountable types of poor judgments.

Changing to an AI system can be scary. We might not want a machine to explain things to us, or to decide things for us. There are legitimate reasons, both instinctive and rational, to resist the insertion of AI in cognitive tasks traditionally performed by humans: voting, deciding guilt or innocence, writing laws. But don't think those preferences will be stable over time. People initially won't accept a lot of the possibilities of how AI can be used in democratic contexts, yet may change their minds with increased familiarity and trust, just as acceptance grew for past controversial technologies like security cameras, credit cards,[2] and even the automobile.[3]

There are good reasons to believe that we will become increasingly inured to AI-infused systems. First: because it's already happening. Many software products, services, and websites that you use have added AI-enabled features over the past few years. Mapping software uses AI to find you the fastest route. Social networks use AI to decide what content to show you. Camera apps use AI to automatically retouch the photos you take.

Second: because we will likely become more comfortable with AI-enabled tools. Industry and academic researchers eager to have their creations used (and monetized) have spent years studying the factors that drive adoption of new technologies. It is no great surprise what works here: demonstrating transparency, compatibility with other tools, reliability, and the ability to simplify the tasks of daily life.[4] Sometimes just signaling these qualities is enough; trust can be a matter of marketing as much as engineering. None of that means that you must use AI for any given task, but it explains why you'll see the people around you becoming increasingly acclimatized to it.

Third: because there are contexts where AI will work well, and consumers will want to use it. Anyone who interacts with people who speak different languages will be an eager adopter of high-quality real-time machine translation. Anyone who must perform a repetitive task that a machine can do almost as well may be glad to save time using AI—unless it takes their job away.

But in the end, it might not matter what we think. Once someone powerful—a legislator, a political partisan, a judge—believes that AI is useful to them, they will use it. They won't ask for permission. Neither will corporations; they can often decide if and when you will use AI in their products or if it will be used on your data. So while earning trust and demonstrating value will be key to making AI stick, that trust and value doesn't need to be equally applicable to everyone.

AI is less a sui generis technology than an accelerant of tendencies with which we are already more or less familiar. Traditional statistical techniques have now been rebranded as AI. This is partly just hype and partly a reimagining of their use cases. That reimagining is important because people's choices in how to use technology dictate its impacts every bit as much as its capabilities.

There are myriad applications of AI in the democratic processes of government. We are advocating for AI that both serves the public interest and advances the goals of our democracy. That starts with understanding all the different processes and systems that make up our democracy and how AI will soon impact each one.

4

AI'S CORE CAPABILITIES

AI will be used in ways that affect democracy, despite—
and sometimes because of—its limitations. AI will often
not be as effective as a human doing the same job. It won't
always know more, or be more accurate. And it definitely
won't always be fairer, or more reliable. But it can have
advantages over humans in four dimensions: speed, scale,
scope, and sophistication. AI's biggest use cases have come,
and will continue to come, in contexts where improvements in one or more of those dimensions make qualitative
differences.

First, consider **speed**. There are tasks that humans are perfectly good at, but they're not fast enough. Here, AI may be
used, even if it is inferior in quality. One example is restoring
or upscaling images, where you have a pixelated, noisy, or
blurry image and want a crisper and higher resolution version. Humans are good at this; given the right digital tools
and enough time, they can fill in fine details. But they are
too slow to efficiently process large images or videos. AI
models can do the same job blazingly fast, a capability with
important industrial applications. AI is used to enhance satellite and remote sensing data, to compress video files to save

on storage and bandwidth,[1] to make video games run better with cheaper hardware and less energy,[2] to help robots make the right movements when their sensor streams are unreliable,[3] and to model turbulence to help build better engines.[4] The speed of AI enables these widespread uses for tasks in which humans cannot keep up.

The second dimension is **scale**. AI will increasingly be used in tasks that humans can do well in one place at a time, but that AI can do in millions of places simultaneously. A familiar example is personalization, as in ad targeting. Human marketers can collect data and predict what types of people will be interested in certain products, or will respond to certain advertisements. This capability is important commercially; advertising is a trillion-dollar market. AI models can do this for every single product, TV show, website, or web user. This is how the modern ad tech industry works. Real-time bidding markets price the display ads that appear alongside the websites you visit, and advertisers use models like this to decide when they want to pay that price— thousands of times per second.[5]

Next is **scope**. AI can be advantageous when it does more things than any one person could, even when a human might do better at any one of those tasks. Generative AI systems like ChatGPT can engage in conversation on any topic, write an essay espousing any position, create poetry in any style and language, write computer code in any programming language, and more. These models may not be superior to skilled humans at any one of these things, but no single human could outperform top-tier models across them all. Organizations will continue relying on human specialists to write the best code and the best persuasive text, but will increasingly be satisfied with AI when they just need a pretty good version of both. And human specialists will increasingly benefit from AI help, too.

Finally, **sophistication**. AIs can consider more factors in their decisions than humans can, and this can endow them with superhuman performance on specialized tasks. Computers have long been used to keep track of a multiplicity of factors that compound on each other and interact in ways more complex than a human could trace. The 1990s chess-playing computer systems like Deep Blue succeeded by thinking a dozen or more moves ahead.[6] Modern AI systems use a radically different approach: deep learning systems built from many-layered neural networks take account of complex interactions—often many billions— among many factors. Neural networks now power the best chess-playing models, and most other AI systems. This is not the only domain where eschewing conventional rules and formal logic in favor of highly sophisticated and inscrutable systems has generated progress. The stunning advance of AlphaFold2, the AI model of structural biology whose creators Demis Hassabis and John Jumper were recognized with the Nobel Prize in Chemistry in 2024, is another example. This breakthrough replaced traditional physics-based systems for predicting how sequences of amino acids would fold into three dimensional shapes with a 93 million parameter model that is naive to physical laws.[7] No one likes the enigmatic nature of these AI systems, and scientists are eager to understand better how they work. But it is undeniable that the sophistication of AI is providing value to scientists, and its use across scientific fields has grown exponentially in recent years.

When an AI takes over a human task, the task can change. Sometimes the AI is just doing things differently. Other times, those changes in degree result in the AI doing different things. In other words, changes in degree may transform into changes in kind. These changes bring both new opportunities and new risks. When AI performs political polling, it could develop a process of listening to complex viewpoints instead of asking simplified questions in a static questionnaire. When AI is used as an arbiter, dispute resolution becomes faster and can be done more proactively, potentially warding off escalations of conflict.

If you're steeped in the technologies of AI, you may have noticed that we've invoked a wide variety of AI types already. That's very much our intent: All these technologies, and ones not yet invented, will be important. We're not going to dwell on the details of how these AI systems work, or their specific advantages and limitations, because the field is changing very fast.

Some believe that AI will become much more capable in the coming decades, perhaps superseding all human cognitive skills. Many AI labs are explicitly pursuing this goal of superintelligent AI, and many science fiction authors have dreamed up both utopian and dystopian scenarios that may result. Whether or not you think this is likely, it's not necessary to assume this kind of extraordinary advancement to see a multitude of ways that AI will influence democracy in the years to come. In this book, we will primarily discuss existing capabilities of AI and improvements that seem plausible within the next few years.

The core strengths of AI point to the places that AI can have a positive impact. When a preexisting system is bottlenecked by speed, scale, scope, or sophistication, or when one of these factors poses a real barrier to being able to accomplish a goal, it makes sense to think about how AI could help. Equally, when these factors are not a primary barrier, it doesn't make sense to use AI.

5

DEMOCRACY AS AN INFORMATION SYSTEM

Democracy has many different purposes: peaceful transition of power, majority rule, fair decision-making, better outcomes. But for the purpose of this book, think of it as an information system for turning individual preferences into group policy decisions, and then executing those decisions through society.[1]

In its ideal form, democracy is a way to balance those group policies and individual preferences fairly.[2] Information distribution and processing technologies, from *kleroteria* to telecommunications, are integral to democracy. AI, too, is fundamentally an information-processing technology. AI takes data as input and has the capability to output predictions, decisions, actions, even new information. To the extent that AI will pervade how all of us consume, reason about, and communicate information in the future, it's inevitable that it will influence, or be integrated in, democratic systems.

Representative democracy is a particular system for managing information flows. Representatives, selected by the citizenry in elections, amalgamate information from large groups of constituents, and then use that information to act on the behalf of those constituents in the decision-making process. Other processes collect information about what people think and want, and how well policies are working: opinion polls, town hall events, public comments, advocacy, lobbying, protesting. Government officials, either elected or appointed, represent citizens in many different decision-making processes. And many more government officials, including administrators, law-enforcement officers, and military officers, represent citizens in another set of processes that executes those decisions. This process happens at a variety of different scales—from towns to nations—and repeats itself throughout the lifetime of any democracy.

The root of many problems in modern democracy is a lack of information-processing capacity. In recent decades, important information distribution and processing channels like unions, fraternal organizations, and religious congregations have been supplanted by digital channels like social media that tend to incentivize decidedly nonproductive civic engagement. Many governments don't have enough humans to do all the work that needs to be done, which creates inequities. Many people have inadequate access to services. Legislatures are not responsive enough to societal needs, and both law enforcement and courts often make decisions based on incomplete and biased information. The gap worsens as society becomes more complex.

Current AI systems are capable of synthesizing, describing, predicting, and deciding. All those capabilities infuse the information system that is democracy. This implies that as AI becomes more capable, either at doing things humans can already do or doing things that humans cannot do, it will have a significant effect on democracy.

The odd thing about this is that democracy is all about putting power in the hands of people, yet the systems of democracy—elections, representation, bureaucracy—are about mechanizing the translation of their preferences into policy and action. These systems make democracy more efficient, but can also make it feel less human. AI can exacerbate that concern, because it can accelerate systemization.

One of the most important information flows in democracy is legitimacy, which is the conception of constituents that a government has rightful authority. Citizens of democracies grant legitimacy to government officials, laws, and policies through electoral consent. This legitimacy has an extraordinary effect: It gives people faith in the value of paper money, convinces them to hand over hard-earned income in the form of taxes, and grants government officials permission to use violence against both individuals and other states. In turn, institutions confer legitimacy on the decisions and actions taken by officials on behalf of the people. Legitimacy is related to, but encompasses more than, trust; we can recognize a politician as legitimately elected even if we distrust them.

As AI becomes more embedded in democratic processes, governments will need to decide to what extent to confer legitimacy to AI tools. They must determine whether to permit AI in political campaigning, encourage the use of AI in public services, and adopt AI to enhance the efficiency of government operations. In well-functioning democracies, citizens generally respect the legitimacy of these government decisions, even when they disagree—up to a point. When governments breach the public's trust through corruption, unjust decisions, or excessive bureaucracy, they sacrifice legitimacy and risk backlash, protest, or even revolution. AI systems that are developed or used in corrupt, unjust, or overreaching ways face the same risks of undermining public trust.

The decision to use AI is very different among authoritarian versus democratic regimes. Authoritarians can much more easily exploit data for AI training and experiment with automation via AI, regardless of concerns about privacy, individual ownership, or due process.[3] They do not rely on the legitimacy conferred by public consent and therefore do not need to build public trust in how they develop or apply AI. This makes them less sensitive to the public harms that misuse of AI may cause. For many authoritarians, the ability of AI to automate visual surveillance, centralize decisions about policing, and monitor and repress dissent is a motivation for, not a risk of, AI adoption.[4]

In extreme cases, AI may undermine the role of legitimacy in democracy. If future AI capabilities grant governments the power to take action (e.g., to administer systems and police public behavior) and apply force (think drones, military robots, and guided missiles) without mass human cooperation, then the legitimacy conferred by the public could cease to matter.[5] This is an alarming idea, but familiar from past technologies. Industrial machines undermined the power of labor relative to capital. Machine guns and nuclear weapons changed the scale at which nations needed to recruit armies to apply force. New technologies routinely put more decision-making power in a few hands, enabling the concentration of power and diminishing the legitimizing potential of popular consent.

Regardless of how you feel about either democracy or AI, AI will have a major impact on democracy because democracy is an information system and AI is a transformational information-processing technology. And it will touch all aspects of democracy: not just elections but lawmaking, the day-to-day functioning of government, litigation and law enforcement, and the responsibilities of the public as well. There are information-processing vacuums in many of these systems, and we already see AI being used to fill the gaps.

But *how* AI will affect democracy is not inevitable. Whether the changes are positive or negative are contingent somewhat on how AI's capabilities evolve, but more so on how we adapt our systems of government and culture over time and the choices that citizens in democracies make. We will take up all these topics as we examine each branch of democratic government, leading to recommendations in Part VII about how to make this work out for the better.

6

WILL AI MAKE DEMOCRACY BETTER?

People are using AI in democratic contexts where its capabilities are poorly suited, where its use is ethically dubious, or even where the notion of automation is offensive to modern conceptions of citizenship and democracy. For example, facial recognition AI has been used in policing before being vetted for racial bias, and predictive models have been used to make decisions about your freedom in legal proceedings before resolving the ethics of that use. In the future, you may even delegate your vote to an AI. It is your prerogative to object to any of these uses. Because these ideas are being tested, it's important to understand their possible ramifications.

We don't know how AI's transformation of democracy will turn out. But we are opinionated about how AI should be used and the way we'd like to see things go. We want AI to be used to distribute access to political power widely, to make policymaking more deliberative, and to center and enhance human civic participation.

There are real risks associated with automating processes of governance and using AI in democracy. The first is **exacerbating injustice**. As an automating technology, AI affords governments the capability to increase the speed, scale, scope, and sophistication of bias, discrimination, and exploitation. It can also be used in ways that help mitigate those injustices, but only if it is used responsibly and if the AI system itself is trustworthy.

A second risk is **inadequate trustworthiness**. Every AI application in this book requires trust. Sometimes only the user of the AI needs to trust it, as in the case of an AI personal assistant. Sometimes society at large needs to trust it: Think of an AI benefits administrator. This entails issues of fairness, accuracy, integrity, privacy, and security. We will discuss applications where people will need to open their personal lives to an AI; they will need to trust that their information won't be used against them, or leaked, or stolen. Performance also matters. For an AI to be trusted in critical applications, it must be reliable.

The third risk is **concentration of power**, another central theme of this book. The race to control AI is increasingly entangling political and economic power in a way reminiscent of the military-industrial complex that emerged in the twentieth century. We will discuss AI use cases like giving corporate lobbyists new tools to write legislative loopholes, enabling demagogues to spew lies tailored to each of us and at scale, and automating systems of oppressive law enforcement. These would all have the effect of increasing the power of the already powerful, and of exacerbating injustice.

These three factors demonstrate how AI can heighten the most pressing threats to pluralistic democracy today and drive our concern that AI is making democracy worse. But we urge against fatalism: Citizens of democracies can grab the reins and steer in a new direction if they choose to do so.

In the final part of the book, we argue for changing how AI is developed and used, in order to achieve a trustworthy system that benefits democracy. We outline what we need to realize this: reform of the corporate-dominated AI ecosystem, resistance to authoritarian and unethical uses of AI, responsible uses of AI where it can safely make a difference, and renovation of democratic systems to mitigate the vulnerabilities AI would exacerbate.

Reform is necessary because the fundamental source of these risks is that AI is being developed by the already powerful, and those individuals and organizations have every incentive to design and deploy it to their benefit. The Big Tech megacorporations touting AI innovations to their investors and authoritarian governments competing in a global arms race for AI supremacy are not doing it because they think this technology will be a boon to government transparency, citizen engagement, or efficient court systems. Nonetheless, automation really could have those benefits, if we could trust the AI systems in those roles and the human-led systems that build and implement them. Trust, however, will remain elusive if the AI system is being built for someone else's benefit.

Resistance is required where the use of AI has anti-democratic impacts. We recognize that much current AI technology is extractive in nature: helping an elite few profit from the creative work of other people, often without compensation, and with little concern for environmental impacts or other societal consequences. As Baroness Beeban Kidron said to the UK House of Lords in 2025, "The spectre of AI does nothing for growth if it gives away what we own so that we can rent from it what it makes."[1]

Yet, there are places where **responsible use** of AI can enhance democracy. We are not techno-solutionists, fabulists, or revolutionaries. We don't regard AI, or any technology, as a primary solution to democracy's problems. And we recognize that implementing new technologies, even when the capabilities are there, is hard. We will give many examples of real-world applications of AI to democratic processes in this book. Those that are being pursued responsibly are largely either (1) the result of an arduous, uphill battle by many people or (2) a work still in progress, with ample opportunity for failure.

Lastly, democracies need **renovation**. Tendencies to concentrate power can be resisted by democratic means. This does not require that we invent new technologies or democratic systems; we already know how to distribute power in a democracy. AI should spur us to adopt reforms that are responsive to the problems it exacerbates, not to search for novel solutions specific to AI. In addition to keeping our eye on how AI might concentrate power, we must spend time figuring out how to shape AI and improve governments to make society more equitable. We'll call out responsive reforms throughout this book.

Regardless of how we feel about AI, we are optimists about democracy. We recognize rising authoritarianism and we see plenty of flaws and failings in how the world's democracies have grappled with technology. Yet we are confident that the best way to address those threats and shortcomings is through democratic systems—through the deliberative power they offer and the legitimacy they confer. We are believers in democratic societies' ability to put technology to work in beneficial ways and to adapt and repair when things go wrong, at least in the long run. We are hopeful that advocates and organizers, and not just authoritarian governments and self-interested politicians, can augment their own power using AI. Every faction in a democracy can make use of AI to help push for the needs of the constituents they represent.

The public interest does not always win out in a democracy, but we are sure that it is more likely to do so when the public is mobilized and engaged. These risks and realizations lead us to the recommendations we urge in our conclusion. Democracy can be made better with AI, for all its risks and limitations, if we do four things: implement reforms to distribute control of AI and steer its ecosystem in pro-democratic directions, resist inappropriate and autocratic uses of technology, leverage AI in responsible ways where it's fit for purpose throughout government, and renovate our democratic systems in ways responsive to the threats and challenges of AI.

Whether AI results in more and better democracy, or less and worse, is up to us. Although the progress of technology might seem inevitable, all constituents have agency in this outcome. And in democracies where that agency is unjustly repressed, we should think about how to use AI in the fight to build more power, not resign ourselves to the idea that it will only be used against us. We can make decisions today about how to use AI that will impact the future of our democracy, through our votes and advocacy. We can support candidates who advance responsible, ethical, and pro-democratic uses of AI to boost democracy. We can

leverage AI in productive ways to increase our own participation and engagement with democracy. And we can reject
inhumane, unethical, and anti-democratic uses of AI—if
we know how to spot them and understand their ramifications. Let's begin building intuition about how to recognize the positive and negative uses of AI, and what to do
about them.

II

AI-INFUSED POLITICS

We'll start our journey into the world of AI and democracy with politics. In democracies throughout the world, politicians need to articulate messages to their constituents, understand what issues their voters care about, coordinate volunteers, and raise resources to support their candidacies and their movements. These are all aspects of information systems, and can derive benefits from AI.

AI-infused politics does not mean robots holding political office, but rather human politicians using AI. The technology will transform candidates' approaches and capabilities in ways similar to how other technologies, like television or social media, have changed what it means to seek and hold political office.

We see political applications as a harbinger use of AI in democracy because none of it requires anyone's approval. No laws need to change, no institutions need to decide to adopt it. All it takes is an individual candidate, or member of their staff, to want to use it. This is already happening in national and local campaigns worldwide. While democracy is more than just electoral politics, the political arena will be a crucible for the impacts AI will have on democracy and its information flows.

But before that, we have the first of our series of background chapters, about the accuracy of AI. What happens when an AI makes a mistake?

7

BACKGROUND: MAKING MISTAKES

Humans make mistakes all the time. All of us do, every day, in tasks both new and routine. Some of our mistakes are minor and some are catastrophic. Mistakes can break trust with our friends, lose the confidence of our bosses, and sometimes be the difference between life and death.

AIs make mistakes, too, and—if we entrust them with important decisions—the consequences can be equally weighty. The difference is that human mistakes are familiar and AI mistakes can be weird and nonintuitive.

Over the millennia, we have created systems to deal with the sorts of mistakes humans commonly make. Casinos rotate their dealers regularly, because they begin to make mistakes if they do the same task for too long. Hospital personnel write on limbs before surgery so that doctors operate on the correct body part, and they count surgical instruments to make sure none were left inside the body. From copyediting to double-entry bookkeeping to appellate courts, we humans excel at mitigating human mistakes.

Life experience makes it fairly easy for each of us to guess the type of mistakes humans will make, and when and where they will make them. Human errors tend to occur at the edges of someone's knowledge; most of us would make mistakes solving calculus problems. We expect human mistakes to be clustered: a single calculus mistake is likely to be accompanied by others. We expect the frequency of mistakes to wax and wane, predictably depending on factors such as fatigue and distraction. And mistakes are often accompanied by ignorance: Someone who makes calculus mistakes is also likely to respond "I don't know" to calculus-related questions.

To the extent that AIs make similar mistakes to those made by humans, those mitigation systems can still apply. It's only when the AIs start making mistakes in new ways—more often, more catastrophically, or just plain differently—that we have to worry.

The current AI models—particularly chat-based generative AI systems—make mistakes differently. AI errors come at seemingly random times on seemingly random topics. A model might be equally likely to make a mistake on a calculus question as it is to propose that cabbages eat goats. And AI mistakes aren't always accompanied by signs of ignorance. An AI may be just as confident when saying something completely wrong—and obviously so, to a human—as it will be when saying something true.

The seemingly random inconsistency of AIs makes it hard to trust their reasoning in complex, multistep problems. (AI companies are investing heavily in this capability, so expect improvements over the next few years.) If you want to use an AI model to help with a business problem, it's not enough to see that it understands what factors make a product profitable; you need to be sure it won't forget what money is.

Both humans and AIs make things up. With AIs, the term of art for these errors is "hallucinations." Confabulations is probably a better word. Both humans and AIs will authoritatively fill gaps of knowledge with fiction. Humans sometimes simply misremember or get confused, and sometimes confabulate deliberately. They might exceed their knowledge in an earnest attempt to help, or try to appear smart in order to protect their reputation, or lie to serve their own interests.

There's no need to anthropomorphize the reasons why AIs make things up, but we must grapple with the fact that they do. This is a major barrier in using generative AI systems in any application where accuracy is critical. AI systems are notorious for making basic commonsense mistakes, but this is improving. AI chatbots are trained primarily to spit out words that make sense together. They have no formal knowledge of mathematics, or how calendars work, or any inherent understanding of temperature. This is quite different than with most humans; few who excel at computer programming and language translation lack the ability to consult a calendar.

But the latest chatbots can command tools. They can, figuratively, pull out a calculator, look up the date, and even check weather forecasts. This is unlike our experience with humans. Most people don't gain entire new categories of skills instantaneously, and certainly not as a group. But software integrations and new models can endow widely deployed AI systems with entirely new capabilities immediately. Of course, that also opens up new possibilities for mistakes or missteps.

AIs can make mistakes because of biases in their training data. Early facial recognition AIs made more mistakes with dark-skinned faces because their training data consisted primarily of white faces.[1] Early speech recognition AIs made more mistakes with high-pitched voices because their train-ing data consisted primarily of male voices.[2] Generative AIs can be racist, sexist, or antisemitic if the texts they are trained with have those characteristics. The core problem is that human failings transfer to the AI. This goes for lying, as well: AI will willingly parrot humans who deliberately introduce falsehoods in AI training data or instructions. One study even found that deliberately training an AI to deceive its users in a particular scenario caused it to act deceptively more broadly.[3]

Mistakes are more tolerable in some circumstances than in others. If an AI used to summarize a news story declares that the US–Mexican border is to the north of the US, it's merely annoying. If a political candidate's AI avatar says that their country should nuke New Zealand, the mistake becomes, at minimum, an embarrassing gaffe. If an AI assist-ing someone with their legal paperwork makes a mistake, it could ruin their life. We should have a very low tolerance for error for an AI that returns wrong answers on an immigra-tion questionnaire, or files a police report spuriously accus-ing a person of a crime, just as we should have low tolerance for error among humans performing those tasks.

This speaks to the importance of knowing what level of accuracy is good enough for each application. And this affects the type of AI we might use in different situations. A primitive US Citizenship and Immigration Services chatbot launched in 2015 returned mostly canned responses and rel-evant links from the agency's website based on keywords.[4] It's very accurate, but inflexible. That's probably the right AI for this type of job, for now. Better for it to be unhelpful than wrong.

In addition to the severity of mistakes, we must consider who they impact. We can recognize that it is unavoidable for an AI tasked with filling out legal paperwork to occasionally make mistakes, and accept an error rate comparable to or better than a human paralegal. But we cannot accept the same AI system if its mistakes are disproportionately concentrated on paperwork from people of a particular racial group.

Some of the same tricks that work to keep humans accurate and honest will work for machines. Forcing people and machines to cite their sources helps, both because it constrains them to make claims that are supportable with evidence and because it offers a route to validating those claims. That said, citations are only as good as the quality of the sources. Misinformation in still leads to misinformation out.

Having another knowledgeable person, or machine, check the work of a colleague helps, too. This is why scientists submit their work to peer reviewers and writers have editors and fact-checkers. Some of the most interesting recent work in improving the factual accuracy of AI tools addresses this: Training specialized AI models on fewer, higher quality sources[5] and using multiple AI agents to criticize and synthesize each other's work product[6] improves their reliability.

In some applications, AIs need to be vigilantly watched, and mistakes quickly corrected. The most obvious examples are modern AI-assisted cars that require human drivers to constantly pay attention and take over the controls in less than a second. That can be very hard to do reliably, and mistakes can be fatal.

In other applications, a human mistake-correcting system can take its time. The work of an AI that drafts news articles, or political fundraising emails, or drafts legislation can be revised by a human editor later. Here, the AI can be viewed as a junior employee. The editing process precedes publication and is faster than a reporter writing a draft, so the AI system saves time and labor.

Putting this all together, we can sketch out a system of phased AI rollout with human oversight. Imagine an AI that is tasked with assigning a benefit or opportunity: eligibility for a government program, a bank loan, or a job interview. First, the AI is only allowed to make easy "yes" decisions. This means that the only errors are that someone is a little better off than they would have been, and the institution using the AI should be obligated to stand by the decision. It's not ideal, but no one is being unjustly denied. Once the error rate on "yes" decisions is at an acceptable level, the AI is allowed to make easy "no" decisions, with a very low tolerance for error. For anything not easy, the AI recommendation is automatically sent to a human for detailed review. Additionally, humans would randomly sample AI decisions for additional review, to be sure that error rates are still low. Moreover, the recipients of AI decisions would have the right to appeal those decisions to a human, under limitations and procedures similar to traditional administrative appeals processes.

In this example, AIs provide speed and humans step in for the difficult decisions. This allows for constant training and tweaking. If a human reviewer regularly reaches the same verdict as the AI's recommendation, then more cases can be classified as easy. If the human reviewer keeps overruling the AI, then the AI will be assigned fewer and easier cases. The AI can keep training on the decisions made by human experts, growing incrementally better. This oversight system allows us to reap the benefit of AI while maintaining human control. It feels much better to have instant denial by an AI, followed by the ability to appeal, than to wait years for a human decision. Here, speed makes the difference.

This scenario is not far off from Canada's 2019 Directive on Automated Decision Making.[7] It requires automated models to provide meaningful explanations of their decisions whenever a benefit is denied or a regulatory action is taken, mandates processes for monitoring the outcomes of automated decisions, and requires methods of recourse to be provided to those affected.

Neither perfect humans nor perfect AIs are possible. In many cases, our choice is not between AI and human; it's between AI and nothing, because there aren't enough skilled humans; or governments choose not to spend money to hire them. Automating claims processing can be superior to consigning people to suffer while waiting for assistance. An AI that summarizes government meetings can be better than having no summarization at all. We must harness our demand for accuracy to continuously improve AI systems, without discounting or ignoring useful functionality.

8

TALKING TO VOTERS

There is a slam-dunk, killer app for AI in democracy to which few would object: automatic language translation. It is tragic when politicians cannot listen to and appeal to every one of their voters due to language barriers, and access to government services and the systems of democracy in a language you can understand is a fundamental matter of justice.

Automatic language translation is changing this. Since at least 2019, the European Parliament has been investigating machine translation to facilitate multilingual debate across its twenty-four official languages[1] and several US states are treating machine translation as the spearhead of their adoption of AI.[2] Immediately after the 2022 Russian invasion of Ukraine, President Volodymyr Zelenskyy coordinated with a Ukrainian company to use AI to dub his speeches into more than twenty languages.[3] Indian Prime Minister Narendra Modi has been using real-time AI translation services to simulcast his speeches to his linguistically diverse constituents since 2023.

Using AI to speak to voters, citizens, and residents across language gaps is just the beginning. AIs can be used to personalize political speech, catering to individual stakeholders.

They can tailor messages, explain concepts, or answer questions differently to you than to your neighbor.

Real-world candidates are already using AI as a communication tool. Argentina's 2023 presidential contest has been called the "first AI election," as both major candidates used AI to develop campaign posters, videos, and other materials.[4] The same year, three US presidential primary candidates—Asa Hutchinson, Dean Phillips, and Francis Suarez—deployed AI-powered chatbots as virtual copies of themselves. In 2024, Pakistan's former prime minister Imran Khan used an AI voice clone to deliver campaign speeches while he was in prison.[5]

AI can help candidates who don't have the money for large human-powered campaign operations. In 2024, a candidate with virtually zero name recognition, Jason Palmer, beat Joe Biden in the American Samoan primary by using AI-generated emails and texts to voters, and an AI avatar on his website that used AI-generated video and voice to interactively respond to voter questions.[6] The result was a fluke—the primary was not seriously contested—but exemplifies the many global experiments in using AI to directly mediate conversations with voters. In the Tokyo gubernatorial race of 2024, the independent, thirty-three-year-old candidate Takahiro Anno managed to come in fifth out of fifty-six candidates[7] after using an AI avatar to respond to 8,600 questions from voters online.[8] Also in 2024, US congressional candidate Shamaine Daniels used an interactive AI robocaller (an automatic dialing device that connects people to an AI chatbot over the phone) to answer voters' questions about the issues in the race by phone[9] and Democrats deployed chatbots to persuade young voters nationally on platforms like Discord.[10] Some of these experiments in using AI to communicate with voters will fail, but it is worthwhile to innovate technologies that provide voters with more access to information about candidates, and balance politicians' reach across the electorate.

This is an evolution of a long-standing trend. The broadcast media of the early twentieth century allowed politicians to reach the entire nation, simultaneously. The fragmentation of the media industry and the rise of the web and social media in the early twenty-first century allowed politicians to tap into distinctive segments of their base, offering bespoke messages to the groups most receptive to them.

AI gives candidates, political advocacy organizations, and even nonpartisan organizations the opportunity to take their communications to the next level. The ability to personalize a message to every voter at once, and then converse with them, makes AI different.

Conversation—engaged discourse where the voter thinks and reflects about their position and how it relates to other perspectives—matters because it's personal and individual. It allows people to reflect on their own lived experiences, rather than on a technocratic policy document. The ability to explain esoteric concepts such as a tariff can make the abstract comprehensible. Like a good human teacher, AI can adapt to a user's language, level of understanding, and the specific concepts they found confusing or illuminating.

We don't yet know the extent to which people will use AI chatbots to learn about politics, but there is reason to believe it will become an important communication channel. Today, people read online newspapers, search the web, and scroll social media for political information. With major search engine providers and social platforms racing to integrate and promote AI search and conversation features, we expect that much information seeking like this will be conducted with chatbots in the future.

Conversation can influence people's partisan beliefs.[11] While it's reasonable to expect that a face-to-face conversation with a human expert (if you can find one with time to spare) would be more persuasive than one with an AI chatbot, researchers have demonstrated that interactions with AI models can change people's policy views. One experiment found that participants who wrote an essay about a policy issue with an opinionated AI writing assistant were more likely, when surveyed later, to shift their own view towards that espoused by the AI.[12] Another study found that some subjects who had bought into false conspiracy theories changed their minds after talking to an AI prompted to offer dissuasive evidence, reducing their belief in the conspiracies by an average of 20%.[13] Further research replicating this experiment showed that fact-based, logical dialogue with AI produced the largest effect.[14]

Existing tools for human-to-human persuasive conversation, like door-to-door canvassing, are only mildly effective.[15] The bar AI tools must clear is quite low. Campaigns that don't have time or staff to reach most voters' front doors are likely to use AI in order to reach more constituents with their message.

If used responsibly, the ability to go beyond robocalls and to deliver personalized and useful messages could allow candidates to assemble new coalitions, uniting voters from across the traditional ideological spectrum based on specific shared values and preferences. It could be used to meaningfully talk to infrequent voters, currently considered not worth the expense of soliciting by a human campaign organizer. It could help campaigns be more precise about the content, frequency, and method of communications, resulting in greater openness and voter satisfaction.

Used irresponsibly, AI could turn voters off even more than the technologies that politicians already use to bother voters, like door-to-door canvassing and robocalls. Even worse, personalized chatbots will give demagogues the power to tell each and every voter what they want to hear, regardless of the truth. AI can be used to mislead voters, create propaganda on command, or invent "facts."

These are many observers' greatest fears. But while propaganda has new flavors—fake news, misinformation, disinformation—it has always been a part of our society, affecting local politics, international relations, and everything in-between. Libelous allegations about British loyalist politician Thomas Hutchinson in the early newspapers of revolutionary America led a mob to burn down his house over a political position he didn't hold.[16] Disfavored Russian revolutionaries were airbrushed out of Stalin's propaganda photos a century ago.[17] Humans have been creating fake news since long before AI.

The current spate of AI-generated fakes is only a warm-up to the greater propaganda risk to democracy posed by AI: the personalization of misinformation. Modern conversational AIs are intentionally designed to be people pleasers and, in the hands of a human demagogue, could be unleashed to manipulate the voting public. AI can tell people exactly what they want to hear. They can do so with precision and differentiation, offering custom lies for each of us. Manipulative AIs operated by a campaign could pinpoint our policy preferences and pull the wool over our eyes about how the candidate plans to handle those issues, or even calculate exactly what allegations (real or fabricated) about an opponent will drive us up the wall. In the hands of a demagogue, AI models can propagate lies with little restraint.

Personalization of propaganda is a risk due to its capacity of targeting disinformation, and due to its potential for surveillance and coercion. Israeli Prime Minister Benjamin Netanyahu was an early adopter of chatbots, deploying them to communicate with voters as early as 2019. Netanyahu's Facebook Messenger chatbot asked users to collect information about the political activities of their friends and family, including their phone numbers and whether they had voted.[18] We expect these tactics to become increasingly mainstream. Depending on your trust in the candidate, they may seem like natural extensions of traditional canvassing or totally Orwellian.

AI-generated propaganda differs from its human-generated predecessors over all four capability dimensions. Speed: AIs can produce propaganda—social media posts, essays, speeches, videos of whatever you can imagine—faster than humans ever could. Scale: AIs can disseminate that propaganda to all social media platforms, deploying millions of accounts, posting and commenting and linking at a scale unimaginable with humans. Scope: Propaganda can be simultaneously created on a wide variety of topics, personalized to each user or demographics (and therefore harder to fact-check), and disseminated across multiple platforms and systems. Sophistication: AIs can immediately translate that propaganda into dozens of languages and personalize it to different cultures, subcultures, demographics, or even individuals.

Already, the political discourse essential to democracy is undermined by surreptitious interventions by bot networks on social media. AIs already discuss politics: They voice opinions, argue, lie, and persuade. They are sometimes operated by foreign actors to undermine democratic societies. Platforms, often supported by law enforcement, sometimes shut down fake AI-operated accounts by the thousands,[19] but they are increasingly indistinguishable from human users. What looks like boisterous political debate between people is sometimes bots arguing with other bots.

The wealthy and powerful have a long record of using technology to amplify their political messages, and AI bots give them new ways to do that. AI-enhanced political speech has the potential to transform public debate and public policy on important issues, particularly those for which political rhetoric is not well-rooted in reality. What bots say will increasingly affect what we all experience, and what we believe others believe. Even if they are not spreading falsifiable disinformation, the deployment of bots represents a significant degradation of vital channels of information flow. In modern democracy, perceptions of norms and the attitudes of others influence our own opinions. If bots take up all the oxygen on a platform, they will affect what people think other people think and may also cause people to disengage from deliberation on that platform.

Democracies should not allow Big Tech and political candidates to dictate how these capabilities will be used. Regulating the manner in which AIs talk to voters should be a matter of public policy, but this is quickly becoming one of the most divisive aspects of AI. Open societies with strong free speech protections will resist controls on what candidates can say to voters, either in person or by means of AI. (In the US, for example, it's legally permissible to lie to voters in political ads.[20]) These are fundamental questions of democratic values that each polity will need to decide for itself, and probably revisit continuously for years to come.

In jurisdictions that do not adopt meaningful regulation of AI political "speech," media and civil society will have a critical role to play. They have always mediated communication between voters and politicians, and should adapt how they do so to new technologies. Just as TV networks have hosted candidate debates, online media can build platforms that help voters access campaigns' AI avatars, and serve as their watchdogs.

9

CONDUCTING POLLS

Public opinion polling doesn't work the way you probably think it does. Most people imagine polls as randomized surveys. A pollster opens a phone book, calls every hundredth number or so, questions the person on the other end of the line, and then publishes the tally. They used to work like that. They don't anymore.

Two basic problems have emerged worldwide. First, nonresponse to polls has skyrocketed. Few people complete mailed surveys anymore, landline use has plummeted, cell phones are harder to align with location and demographic data, and fewer people answer their phones in any case. In 1979, University of Michigan pollsters received responses from 72% of people they called for the US Survey of Consumer Attitude; by 2013, it was only 16%. It's only gotten worse since. Second, people don't always tell pollsters what they really think.[1] Maybe they are hiding their true thoughts because they are embarrassed about them. Maybe they are behaving as a partisan and telling the pollster what they think their party wants them to say. Maybe they don't trust the caller's motives.

Pollsters have adapted to these challenges and, remarkably, stayed fairly accurate in spite of them. They use complicated statistical models to weight and rebalance survey data.[2] These models intentionally shift the results based on the surveyor's informed expectations.[3] Pollsters also find it valuable to integrate information from outside their own surveys, such as data on the "fundamentals" of the economy. On top of that are so-called "house effects." Some types of polls systematically arrive at different results than others, even when they ask the same questions.[4] So pollsters adjust for that, too.

The result is that, when pollsters poll people, they take each response as a datapoint and consider it alongside other data. The results are manipulated and massaged. None of this is necessarily malicious or suspicious. Pollsters use models to figure out what precise adjustments to make so that the outputs match reality as best they can. Sometimes these adjustments go sideways: Modern polling isn't perfect, but these changes have kept it useful.

Pollsters have also developed ways to listen without explicitly surveying people. Over the past two decades, "social listening" has emerged as a critical campaign tool. As soon as a news event happens, or a candidate gives a speech or interview, reactions from commentators, activists, and everyday voters are posted to social media. Campaigns use tools to sweep up and process all that data—sometimes millions of individual posts—so that they can learn from it. Large businesses like Brandwatch/Crimson Hexagon and Meltwater have been built around these capabilities. This type of data synthesis is even used to predict election outcomes, like traditional polls.[5]

The natural extension of these applications is to use AI to develop a different type of generalization, across questions. Take a simple example: Suppose you have the results from polls on a ballot question to raise a tax rate. Now suppose the same voters are asked right away about a new proposal to lower that same rate. You can guess the response to the new question based on the last one and your knowledge of how the questions are related: The obvious prediction is that the voters' responses will invert. Now suppose the question is about changing a deduction instead of a tax rate. Suppose that the stock market took a hit since the last vote was taken. Predicting the result is now harder, but the same principles of inference apply.

AI tools can apply this kind of reasoning with instantaneous speed, which will make them intoxicating for politicians and campaigns. AIs can be trained on all the types of data we've discussed: survey data, voting results from past elections, comments from social media. You can condition the AI to respond to questions as if it were an individual voter with a particular set of traits. You can simulate entire polling campaigns, consisting of thousands of individual AI agents, and generate a dataset that looks like a survey of humans.

Of course it's not the same as a human survey, but the AI has the advantage of speed. It will answer instantly, and you can ask it a follow-up question right away. It has greater scale: A thousand campaigns can use it at the same time, even candidates for local offices who don't usually have enough money to run a human poll. It has great scope, since it is able to answer any question, even if that question has never been the subject of a survey before. And, like the models we use to weight and adjust surveys today, it has a high degree of sophistication; it can consider many factors that might influence a population's response to a question.

Taken to an extreme, this would be an approach to polling with a fundamentally conventional bias. As the ethicist Shannon Vallor has written, AI models aligned to historical data are effectively stuck in the past.[6] Would AI have anticipated the twenty-first-century global shift towards greater support of LGBT rights, for example? If we use AIs to guide political decisions, they could stifle our imagination about progress and change. The lesson here is not to use AI models to try to dictate a polity's future, but rather to help synthesize its current and past preferences. This applies to human surveys as well. A survey only tells you what people thought of a question when you asked it, not how they will feel as events change, as they grow and learn, and as the world evolves.

AI predictions are still inferior to human surveys, and campaigns are unlikely to trust them as much. But some will use them anyway. AI agents always pick up the phone, so to speak, no matter how many times you call them. They will answer thousands, or millions, of questions. They will answer open-ended questions. A political candidate or strategist can ask an AI whether voters will support them if they take position A versus B, or endless tweaks of those options. It's like holding a million focus groups: They can direct questions to simulated married male voters of retirement age in rural districts without college degrees who lost a job during the last recession. If they have a secure AI, they can ask volatile or even offensive questions without fear of them leaking to the real electorate: "Would you be more or less likely to vote for me if I flipped my position on immigration? What if I started/stopped calling global warming a hoax?" The result could be much more informative than a survey of humans could ever be.

An AI model will never preemptively guess your opinion better than a pollster who directly asks you a question, but their guesses will improve over time. The trick is to give them more context. Newer models are increasingly trained with current data and are able to search the web or dynamically query databases of news and information for up-to-the-minute detail. Increasingly, AI models are like humans, swimming in a roiling sea of continuously flowing information. This will help them respond more like us.

By walking along a path of gradual changes, polling will arrive at something that seems radical: asking computers to tell us what humans think without consulting the humans. Although this probably sounds like a terrible idea, it's already happening.

Many of these developments are being tested in the field of market research. Companies are willing to pay a lot to learn what consumers think, and they have the same problems with traditional polling that politicians do. The company Expected Parrot has built tools for creating AI personas that respond to survey questions exactly as described above.[7] Several other research groups have been studying the potential for AI-based polls to be biased, or to exaggerate the prevalence of extremist views.[8] Our own research group at Harvard has published demonstrations of using ChatGPT to simulate outcomes of the US Cooperative Election Study survey.[9] It worked well for predicting what different groups think about abortion, but flubbed a question about American support for intervention in the war in Ukraine. The simulated personas don't perfectly mimic individual human responses, but the results are enticing to those looking for tools to rebuild a decaying infrastructure for public opinion research.[10]

AI will work best to augment rather than replace human polls. Researchers have shown that using AI to augment human surveys can maintain accuracy while decreasing the number of human responses needed, saving costs and time.[11] AIs will continue to grow in sophistication for interpreting data, as will their utility for extrapolating from what was asked directly to what people think about the underlying issues. AI will be used to identify when polls will be most wrong or uncertain. They will recognize which issues and communities are most in flux, what questions or communities are too poorly represented in online speech to make good inferences about, and where a model's training data is liable to steer it in the wrong direction. In cases like these, future AI models can send up a white flag and indicate that they need to engage human respondents to calibrate against real people's perspectives. They might suggest ways to ask questions that would elicit more useful responses, and ways to reach people who don't usually participate in polls. AI agents can even be programmed to automate surveys conducted by humans.

We already live in a world of human–AI polling chimeras. Today, humans complete surveys and computers fill in the gaps, because we need algorithms to tell us what all of this messy human data means. In the future, the opposite will occur: AIs will fill out the surveys and, when unsure what box to check, humans will fill in the gaps. This isn't so radical after all: It's AI doing familiar things differently, not really AI doing different things.

10

ORGANIZING A POLITICAL CAMPAIGN

Electoral campaigns are often embodied by one person: the candidate. But effective campaigns are acts of coordination, sometimes on a massive scale spanning the largest nations on Earth. A candidate relies on the combined efforts of campaign managers, campaign staff, advisors, donors, consultants, volunteers, party officials, and so many others. The campaign only works when they are all performing well, and in concert with each other.

Technology already plays a massive part in enabling modern campaigns. National campaigns typically use centralized software to distribute door knocking, phone calling, and texting lists to armies of far-flung volunteers and to collect data from each of them: Did anyone pick up the phone? Did they say they would vote for the candidate? There are opportunities for AI to assist each of these campaign roles, but its real value will come from aiding in their coordination.

In the near future, volunteers will have interactive AI prompts that can both ingest and respond to any kind of nuanced information. A volunteer can report a tough question posed by a voter and receive specific guidance from the AI on how to answer it, pinging the campaign manager if it needs instruction. The AI can recognize if a similar question is popping up across a region, and send out a push notification advisory to volunteers with new instructions.

Campaigns will use AI to augment their human canvassers' efforts to convince supporters to show up for them at the polls. In the US, where voting is not compulsory, it is estimated that in-person get-out-the-vote canvassing contact increases the odds that a voter will turn out to vote by about 6%,[1] far more effectively than the half-percent increase gained from a reminder message on social media.[2] But in-person canvassing requires intensive human resources. For campaigns looking to increase their scale and effectiveness, using AI to personalize those online contacts and make them look and feel like in-person conversations will be very attractive.

There are already providers offering AI-infused coordination apps to campaigns. Votivate, built for the US Democratic Party, uses an AI voice to call voters in multiple languages; produce campaign strategies using information about the race, opponent, district, and current fundraising operations; and to individually tailor campaign videos to voters.[3]

The advantage that AI offers in these contexts is not quality, it's coordination. This is a case where a difference in scale becomes a difference in kind. A single AI will be more consistent than a hundred volunteers, and the ability to coordinate at scale enables new and different campaign strategies.

Campaigns rapidly integrate useful technologies, whether we voters like them or not, and this will likely be true for AI. It probably seemed weird the first time you got a text message from a campaign; now many of us get a few every day.

It might have looked odd the first time you saw a campaign volunteer staring at their phone to get instructions about which house to go to next; now every canvasser goes to the doors an app tells them to.[4] The next time you say goodbye to a door knocker, you might see them dictating whatever you said back to an AI campaign coordinator. Eventually, your whole conversation may be automatically captured by an AI for analysis.

All this AI involvement may make political campaigns, and politicians, more responsive to people. The primary role of campaign staff and volunteers today is to pump out their candidate's message. Campaigns will generally be happy to obtain extra feedback from voters when they can get it, and AI can help with that.

If these tools were to become widely available, many hope running for office would become less expensive, opening campaigning beyond the wealthy and well-connected who already have the means necessary to galvanize support to run for office. However, past technologies like the internet have hardly dented the competitive advantage that wealth and privilege provide to candidates, so there is reason to be skeptical that AI will.

Some of the most interesting dynamics will occur between campaigns. AI will make it easier for candidates to carve out niches and thrive within them. They will help little-known local representatives with "ideologically pure" stances find and attract compatible single-issue voters; this is particularly important in political systems with fragmented parties. They will help nontraditional candidates activate new pockets of citizens who don't usually pay attention to campaigns. These individuals typically won't have the resources to launch a nationwide advertising blitz, but they stand to gain from targeting and tailoring capabilities that position their name and message in front of the most receptive few thousand voters. AI will also influence local ballot initiatives, where people are voting on a specific policy and not for an overall candidate.

All this is likely to make future election cycles more contested, and chaotic. We can expect more candidates trying to reach their pocket of voters with narrowly tailored messages, perhaps too radical to appeal to most voters. And these candidates may survive longer in the campaign and compete across election cycles, since the scalable personalized outreach of AI can help them access additional resources to stay afloat.

The coordination benefits may be particularly useful to international political movements. Campaigns that are active across many jurisdictions, time zones, and languages are challenged to keep many parties moving in sync. Think of human rights and climate campaigns that are seeking to create global change and international cooperation, but face wildly varying cultural contexts, legal systems, and political actors. A multilingual and multicultural AI could help these campaigns localize talking points or even model legislation across borders. Unfortunately, this coordination benefit would equally apply to "astroturfing" campaigns, fake grassroots efforts, and foreign influence operations that will find it easier to activate inauthentic movements across countries.

Sophistication benefits could also accrue. Campaign organizations of all kinds will try to use AI tools to make smart bets about how and where to spend their money. To the extent that AI can make better, faster decisions about where to direct funds, predicated on more context and variables than could be considered by a human political strategist, this kind of automation will create real benefits for the parties that embrace it.

Some coordination problems are common to both humans and AIs. Both can go off script, such as when a reporter asks a politician a question that they should know better than to answer. AI models do this too; there is a cottage industry in discovering unlikely phrases that will prompt generative AI models to forget their core instructions. OpenAI, for example, has struggled to enforce rules against using Chat-GPT for politics.[5] Developers are eliminating those hacks by adding more constraints to AI. Campaigns may expose only the most locked-down interfaces to volunteers, contractors, the media, and the public.

And, as badly as humans if not more so, AIs can fail to understand underlying dynamics. Their judgment may generalize poorly to new situations and questions, unless they are informed by reliable data and smart analysis. Humans and AIs alike will make different judgments about a political issue before and after being shown relevant data, such as opinion polling. The campaigns that benefit most effectively from AI-assisted strategy development and political coordination won't necessarily be the ones with access to the best and most effective apps and most advanced modeling capabilities; they will be the ones who make the best decisions about what information to feed into the AI tools they use, and do the best job of setting goals for them.

Of course, sometimes AIs will get it wrong. Just like humans looking at flawed polling data or making bad assumptions about shifting political winds, AI tools will sometimes pull money from the races that need it most. The AI tools might be faster than humans, might even be smarter than humans, but just like humans, they will be fallible and reliant on the data fed into them.

AI tools can be used as a political shield. Candidates who formerly blamed the intern for a goof-up will have a new scapegoat. Party leaders can pass the buck on tough decisions; a bigwig pulling ad spending from a candidate in a losing race can tell them, "It wasn't me who lost faith in your electoral chances, it was the AI!"

Finally, AI may exacerbate factionalism within democracies. We are already seeing AI become yet another politically polarized issue. In the US, the Biden administration issued executive orders and urged international cooperation restraining AI and applying guardrails to its development and use. In response, the succeeding Trump administration retracted those guardrails, embraced automation of government, and urged the EU and other states to do the same.

But there may be more profound, longer-term effects of AI that change how factions themselves behave. In systems that traditionally have high degrees of federalism and independence, the coordination capabilities of AI may induce greater uniformity within political parties. It's harder to deviate from the party platform if leadership can instantaneously dictate political messaging to every candidate using AI. If AI becomes highly effective at persuasion, it may lead parties to broaden their outreach, chasing voters traditionally affiliated with their competitors. If not, it may further parties' tendencies to cater to their own base, focusing on messaging to and activating those that already agree with them. At the same time, if AI becomes an effective tool to inform voters about the nuances of the issues they care about, it may decrease the signaling power of political parties—the party label next to a candidate's name on the ballot means less if voters are more familiar with their individual positions and qualifications.

11

FUNDRAISING FOR POLITICS

In many democracies, elected officials spend a shocking amount of time asking people for money. US presidential campaigns are multibillion-dollar businesses, and members of Congress spend most of their time fundraising. Even at the US state level, where the total cost of campaigning is much lower, the portion of a candidate's time dedicated to fundraising can be surprisingly high.

The US has particularly pernicious political finance problems, but exorbitant costs are not a uniquely American issue. India spends even more on elections than the US.[1] Funding in other democracies is smaller, but growing. In 2023, the UK nearly doubled its parliamentary election spending limits to £35 million per party.[2] Dynamics are different in different countries. Some, like Australia and Indonesia, seem to be evolving heavily digital political fundraising systems.[3] Brazil's elections were dominated by corporate funding prior to a 2015 court ruling, which has had limited effect at reducing corporate influence.[4]

One consistent worldwide trend is an increase in small donations. This is attractive to political parties because it gives them a new call to action to stoke political engagement.

And, beyond political parties, it gives outside-the-mainstream movements a mechanism to organize and garner resources.

AI will change fundraising dynamics in ways large and small. People are already experimenting with letting autonomous AI agents loose to try to raise money for charities. Even if these kinds of vertically integrated AI fundraising schemes never work, AI will impact each level of the fundraising stack.[5]

Its most immediate impact has been to automate the process of soliciting funds. The simple, repetitive tasks of picking potential donors from a list, tracking down contact information, and gathering donor profiles are all well within the means of existing AI systems. The content-generating steps of drafting personalized solicitation emails and even dialing for dollars are, too. The US Democratic Party has already found that AI-written fundraising messages may be more effective in soliciting donations than human-written ones.[6] If it was inevitable twenty years ago that campaigns would adopt social media platforms to solicit donations, it is equally inevitable today that they will embrace AI tools.

AI tools may or may not be dramatically better at persuading people to donate than past generations of targeting technology, but they can operate at superhuman scale and sophistication, tailoring and customizing messages to every donor prospect, and A/B testing everything. An AI-powered web display ad next to a news story can tweak ad copy to be directly responsive to the content of the story; it could stoke outrage about something that just happened, or articulate a candidate's disputation or refutation of those events.

For some prospects, especially well-connected, high-dollar donors, AI intermediation won't work. Their reason for contributing to a campaign is to buy access to a political system and people close to power. Those big donors will continue to get the white glove touch, not the more impersonal AI. This creates an obvious inequity, but not a new one.

The small-scale implications of the adoption of AI for political fundraising are clear, and somewhat dispiriting. As campaigns at the local, national, and even international level increasingly enlist AI tools to reduce the cost of solicitations, we can expect more solicitation. If it perturbed you to see postal mailers turn into robocalls and email spam turn into text messages and ads on social media, prepare yourself for a lot more of the same, coming from candidates, parties, and organizations.

There will be ways for average citizens and small dollar donors to fight back, if they wish. Just as previous generations of AI have powered email spam-fighting systems,[7] AI tools will be increasingly deployed to filter your inbox, text messages, and incoming calls. It will be easy enough for citizens of the future, weary of perpetual campaigns, to instruct their virtual assistant to ignore all political ads and solicitations. The tactic of campaigns will then become to convince you to safelist their candidate within your AI spam-blocking tool. This will continue a long history of AI spam-filtering technologies influencing political fundraising. In the 2020 US presidential primary, for example, candidates complained of huge differences in which campaigns' emails made it past Google's AI filters to display in Gmail users' primary inbox.[8] In 2022, Google was sued by the US Republican National Committee for allegedly penalizing its solicitations in its AI spam filter (the lawsuit was dismissed).[9]

AI will create new ways to subvert campaign finance rules that were written with older technologies in mind.

For example, many countries have explicit rules around the use of robocallers to solicit donations. Previous generations of robocallers used push-button soundboards to help foreign contractors speak with the accents of local volunteers, often violating local law and sometimes generating tens of millions of dollars from unwitting donors.[10] Those types of schemes can become dramatically easier to execute with AI.

AI fundraising should prompt democracies to revisit campaign finance regulation. Many democracies need stronger limits on how, from whom, how much, and from what jurisdictions or accounts campaign contributions can be made. For a start, countries should mandate disclosure of AI-generated ad copy, text messages, and calls, so that voters can distinguish mass messaging from individual conversations from personalized advertising. Since 2018, California has mandated that commercial and political advertisers disclose when their communications are made by bots.[11] There will always be fuzzy boundaries between benign use cases like AI spell-check or copyediting and potentially nefarious use cases like disguised and manipulative AI solicitors. The EU AI Act's Article 50 mandates disclosure of AI-generated audio, video, and text with reasonable exceptions, such as when the AI-generated content has been reviewed by a human or is published under human editorial responsibility.[12]

Disclosure is just one way to empower people to decide for themselves what they are comfortable with. Many countries also have Do Not Call or Text lists to help people opt out of spam messages, and yet countries like the US and Canada have exempted political campaigners from obeying them.[13] That should be changed.

Social media companies could choose to restrict these messages, too. Facebook banned political advertisements for about three months after the US 2020 election[14] and many social platforms give users tools to opt out of political advertising, though they often make the process difficult. Platforms should be required to make this easier. Finally, people can take a grassroots approach to monitoring AI-generated advertising. Organizations like ProPublica have launched crowdsourced, international efforts to track online advertising: what the messages are and who they target.[15] These efforts will increasingly be vital to counterbalance the proliferation of AI-powered misrepresentation and manipulation, particularly in jurisdictions like the US that consider money in politics to be constitutionally protected speech largely beyond the reach of government regulation.

For democracies willing to use it, regulation can limit the power-amplifying potential of AI in privately funded campaigns. Candidates should work to win the support and contributions of regular voters, particularly those in their districts, rather than delegate all of that work to a machine.

There are familiar reforms responsive to this challenge. On the campaign finance side, stricter caps on contributions (applied both to candidates' campaigns and outside political organizations) would discourage candidates from catering to billionaires while handing over grassroots organizing to AI. On the political advertising side, stronger rules and rigorous enforcement on disclosure, false representation—as hard as that is to enforce—airtime, and spending limits would limit the disruption that AI fundraising could impart.

While the most visible impact of AI integration into campaign fundraising will be on the proliferation of personalized solicitations, there will be more macroscopic, landscape-shifting effects as well.

AI will provide people with new tools to decide where to direct their giving. Today, many small dollar donors are motivated by causes rather than politicians; they may give to interest groups, parties, or—in the US—a political action committee rather than a political candidate. These organizations research and make decisions about where to distribute funds. AI tools could personalize that service, automatically directing donations based on the donor's values, policy preferences, or political interests. When such tools exist, campaigns will become highly incentivized to pander to the AI, much as candidates curry favor from national parties and interest groups today. Since AI can also increase the coordination of these entities, it could be used to discipline multiple elements in a coalition. A campaign might be less likely to signal compromise to voters from another party if they know their biggest donor has an AI watching everything they say.

All this means that AI may have a dual influence on campaigns. It may lower the barrier to entry, enabling more candidates to run, while simultaneously increasing the advantage of those with the most resources, helping the top end of candidates to consolidate power.

If we can evolve campaign finance regulation to anticipate the capabilities of AI-augmented campaigns, we can set up a system that encourages pro-democratic outcomes. We can widen the base of political participation. And, as we will see in the next chapter, this is not the only way that AI can make political campaigning more accessible.

12

BEING A POLITICIAN

We're not going to elect AIs as political leaders anytime soon. This is not something that voters, political parties, or the legal establishment are seeking—or will accept. Such a development would require major changes to the law. But as we experience an increased level of AI automation and assistance, our human politicians will lean more on AI. This will largely happen invisibly, and we will mostly be okay with it—or, at least, unaware and apathetic. It's just another step along a path we've trodden for a long time.

When a president makes a heartfelt public speech, we might resonate with it emotionally, but deep down we suspect that a speechwriter wrote it. When a legislator proposes a new bill, we might support it enthusiastically, but we assume that they didn't write its text. And when we receive a holiday card in the mail from any of these people, it may make us smile, even if it was signed with an autopen. Those things are so much a part of politics today that we don't even think about them.

Modern politicians are already collective and cyborg-ish. A politician today is really a complex socio-technical system with a single person as the public face. In presidential executive systems like that of the US, we like to pretend we're just voting for one person, but we're really voting for the whole package: a cabinet of allies, a coterie of advisors, a party's policy platform. A good legislator manages a team of experts, takes advantage of research agencies, and learns from written summaries of events and issues. They prompt human speechwriters to write their speeches and human lawyers to draft their bills.

Gathering a good combination of people and tools is a major political skill, one that makes a huge difference in a legislator's ability to govern. There is a singular figurehead: we vote for the individual, inspired by their ideology, charisma, or shared identity, then we hope that the system they bring along with them matches. Integrating AI tools well will become part of this skill.

In 2024, Victor Miller, a mayoral candidate in Cheyenne, Wyoming, pledged to defer all of his decisions to an AI. Standing in front of a campaign sign reading "AI FOR MAYOR," he let an AI take questions at press conferences. That AI responded in real time in a synthesized voice.[1] Miller came in a distant fourth in the race, with 3% of the vote.[2] British businessman Steve Endacott tried to do much the same thing in running as an independent candidate for Parliament in 2024, campaigning on the idea that he was "controlled" by an AI avatar (which his company built) that would solicit feedback from thousands of residents on a weekly basis.[3] He finished last, with 0.3% of the vote.[4]

These examples are extreme, but this is an approach we expect to see others try. It is not unusual for politicians and parties to expressly rely on outside forces to guide their judgment and decision-making. Leaders worldwide tout their deference to religion, or science, or political philosophies. When sincere, that deference is a way of signaling to voters that they will adhere to a certain worldview—one shared by their voters—that provides prescriptions and justifications for their policy choices. We can think of an AI-led political party or politician as something like that: a way to predictably encode a policy platform through our vote. Even voters who have no interest in electing AIs may be attracted to AI as a tool to more reliably constrain their elected officials to adhere to their political interests.

We expect politicians to be early adopters of AI because they are constantly seeking an advantage. Their choice to use AI does not depend on whether their voters trust AI or think political applications are a good idea. For example, candidates will use chatbots to prepare for debates: A nonhuman persuader could serve as a foil to a human rebutter. It will be able to answer questions and spar with their opponents, mimicking their actual political rival's voice, talking points, and speech patterns with a far better impersonation than a typical human stand-in. This has already happened in front of voters: When incumbent representative Don Beyer of Virginia's eighth district declined to participate in one of several candidate forums, two independent challengers debated an AI representing him and trained on his writings.[5]

In the future, we'll accept that politicians rely on AI tools to draft and revise messages, to simulate and understand their electorate, and to help make campaign and policy decisions. We'll soon regard it as normal that most of their communications are written by AI. None of this will necessarily be bad, or change the nature and ideology of political leadership. But it will change the skills necessary to be a politician, devaluing leadership and, even more than today, benefiting charismatic individuals who are basically actors on the political stage.

More generally, an AI may be able to direct or even found a political party. The first example of this to emerge was the Danish Synthetic Party. In 2022, an artist collective in Denmark created an AI chatbot to interact with human members of its community, exploring political ideology through conversation and analysis of historical party platforms in the country. The party failed to receive enough signatures to earn a spot on the ballot.[6]

Future AI-led party formation initiatives may yet succeed. The Synthetic Party demonstrated that an AI can formulate a set of policy positions targeted to build support among people of a specific demographic. An AI could also be designed to attempt to create an effective consensus platform capable of attracting broad-based support. Particularly in a European-style multiparty system, we can imagine a new party with a compelling narrative—and an AI at its core—winning attention, signatures, and votes.

Acceptance of AI use may grow only if there is adequate transparency and disclosure. If you're a voter whose first language is different from most other people in your country, you would probably welcome the use of AI to translate candidates' speech so you can understand their message more easily. You might already use an app on your phone to do this. But if they tried to trick you into thinking they speak your language themselves, they would likely lose your trust.

Like so many other issues, AI will attract political polarization in unexpected ways and by unpredictable events. Today, leftists decry corporate bosses using AI to crush labor, hawks champion AI development in an arms race against foreign foes, conservatives claim AI is biased against them. All this will change over time. Political actors will take up the mantle of high technology and AI for unexpected reasons, and those parties will be more likely to accept and push on the boundaries of AI-assisted politicking.

Our guiding principle here is to welcome uses of AI that distribute power and to resist those that concentrate power. If an opposition candidate running for local office gets help to bootstrap their campaign from an AI speechwriter, that's a good thing. That candidate should still be held fully responsible for the words they speak and the ideas they represent, but they should not be discredited for using the help of machines any more than a better-funded candidate would be criticized for hiring human advisors. Candidates that propose to turn their policy platform over to AI models, however, should be criticized with all the same force that would be due to a candidate that promises to defer decisions to another private interest, like a billionaire or corporation.

III

AI-ENHANCED LEGISLATORS

Once politicians are elected to legislatures, they're supposed to legislate. AI will be part of this process as well.

Like political campaigning, policymaking is rife with complicated information systems that stand to benefit from AI automation. Lawmakers need to be aware of the needs of their diverse constituencies, and to consider complicated political, scientific, and economic implications of policy choices. They need to articulate those choices in arcane legal language that will be embedded within thousands of existing laws. And they need to do so within an adversarial system, full of lobbyists looking to introduce loopholes and opposing factions looking to promote their own priorities and interests.

Using AI for these processes poses both opportunities and risks, for lawmakers and for those who live under the laws they craft. Changing one word in the law of a large country can affect millions of people's lives. This means we need to talk about how AI explains itself, because one of the critical components to trusting AI with any decision is understanding how that decision is made. And when an AI makes a mistake, how do we assign blame?

13

BACKGROUND: EXPLAINING ITSELF

We humans often need explanations. It's not enough for a doctor to offer a diagnosis, or a judge to issue a ruling. To understand and trust experts' decisions, we want to understand their reasoning. It's no different for a political operative proposing a campaign strategy, or a politician proposing a legislative agenda. Don't just tell us what, tell us why.

Much has been written about the dangers posed by the inability of modern AI systems to provide explanations understandable by humans. They can diagnose chest X-rays, or make a recommendation about whether to approve a bank loan, but can't be relied on to explain their reasoning. More so than other technologies, many AI systems are notorious for being "black boxes." Their high degree of sophistication, accounting for huge numbers of variables in complex ways, makes it hard to isolate the discrete impact of any single factor. Even engineers who build AI systems can't explain why they produce the output they do.

This is an area of ongoing research, and we expect to see advances in "explainable" and "interpretable" AI over the next few years. However, there might be theoretical limitations to explainability. Human explanations are a cognitive shorthand, optimized to the way we humans make decisions. AIs don't make decisions in the same manner that humans do, and the number of variables they are capable of considering might be far beyond what humans can reasonably understand. AI decisions simply might not be conducive to human-understandable explanations, and forcing them to produce both decisions and explanations might affect the quality of those decisions. Others argue that forcing AI models to be understandable improves their performance.[1] It probably depends on context.

There are two things we can learn from human experience to understand how and when AIs need to explain themselves. First, human decisions can also be unexplainable. Your brain is a black box; no one can open it and examine how you reached any particular decision. Yes, you might be able to produce a verbal explanation, but human explanations aren't necessarily reliable.[2] Neuroscientists don't know how much of that unreliability is an outgrowth of your predecision reasoning and how much is your post-decision justification. You might not even know how you arrived at a conclusion; if your thought process is more intuitive than analytical, the true explanation may be buried in post hoc rationalizations.

Second, your explanation is often unnecessary. Even though your true intent may be subjective and unknowable, I can judge your decisions independently by observing their effects. Are they correct? Are they fair? Are they racist?

Humans don't want explanations as much as we want consistent justifications or rationales.[3] Justifications are reasons for a decision that are external to the decision-maker, and that are based on the rules and norms of society. We don't subject a loan officer to an MRI machine to determine which regions of their brain light up when they consider loan applications; that kind of causal explanation would be wholly unsatisfying. Rather, we want the ability to verify that the decisions made by the loan officer consistently follow objective criteria established by law, by the lender, and the expectations of our community. Similarly, if an AI were to decide that you don't qualify for a bank loan, you would want to know what could have helped you to get that loan. What if your income were higher? What if you lived in a different zip code? What if you were a different age, or gender, or race?

Some justifications are acceptable. "We decided that it wasn't profitable to lend to you because your credit history shows a pattern of defaulting on loans" is okay. "We threw all the applications up in the air, and offered loans to the ones that landed face up" might be okay, if the system was meant to be random. "We gave loans to the people with European-sounding surnames" is not. We don't want to rely on the loan officer's subjective explanation of their approach, we want to validate that their decisions follow our expectations of fairness.

For AI to be justifiable, there must be access, transparency, and independent evaluation. Research on recidivism prediction systems has demonstrated how justifications for AI models can be built to disguise bias, offering seemingly benign rationales to nominally explain predictions that are actually made based on race.[4]

Many AI developers don't want to provide this sort of counterfactual justifiability. Legitimately, developers may fear that too many explanations would create a road map for those who are trying to game the system. Cynically, it would afford users a more solid basis for criticizing their products, even alleging misconduct. AI companies may want to prevent people from trying a thousand counterfactuals to understand the systems they create. But for many applications, especially those involving democracy, some level of justification or counterfactual explainability will be essential if people are going to trust the output of these systems.

This sort of rational verifiability is an easier goal than robust explanation. A justifiable AI system for loan approvals must be auditable in order to engender and ensure trust. The system could be tested by changing one variable at a time—for example, higher income, lower rent, different zip codes, ages, races, and genders—so that we can ascertain the inputs that would have led to different decisions.

A verification process of this type allows AI systems to be contested. If you can convince a human that an AI made a poorly justified denial of your insurance claim, or determination of your bail, maybe you can persuade the organization in question to overturn the machine's decision. One of the bedrock rights people have in a democracy is the right to contest a government decision. The combination of justifiable AI with systems of appeal, like those we already use to contest human judgments, can help mitigate the potential harms of automated decision-making.

Our perceptions of AI decision-making systems will change over time. The more familiar we are with any given system, the less we demand explanations for its decisions. As the sorts of AI systems we describe in this book become familiar enough and accurate enough, they won't need to routinely offer explanations or justifications. Consider your bank's app for depositing checks, or the navigation app on your smartphone, both of which use AI. Your trust is an outcome of their performance rather than of any explanation of how the AI works. Similarly, consider AI systems that help campaigners pick voters to target for canvassing. If they see other successful candidates using the system, they probably won't demand justifications for each home the AI tells them to visit. Justifications will matter most not to users of the system, but to those impacted by it. When a dozen different campaigns all use AI models to direct canvassers to the same voter's house, that person may be quite interested to know what it is about them that the AI likes so much.

14

BACKGROUND: WHO'S TO BLAME?

AIs will dispense legal advice. That advice will inevitably include mistakes. Someone, somewhere will eventually be harmed by one of those mistakes. Who's to blame? Or, in the context of a litigious society like the US, who gets sued?

The person or company running an AI legal chatbot might be responsible. Alternatively, the organization that created the AI model might be responsible. Or the individual engineers who built the AI might be responsible. Or maybe an entrepreneur licensed a generic AI model, turned it into a dispenser of legal advice, then licensed that product to legal websites; they might bear responsibility for the AI's mistakes. Or maybe the user is responsible for making the decision to trust the AI; that's a convenient scenario for AI developers. Caveat emptor: Let the buyer beware.

Alternatively, we might imagine that the AI itself is responsible. This is the stance that Air Canada took in 2023, when its online chatbot offered a customer a discount it was not supposed to give. The airline argued that the chatbot was a "separate legal entity that is responsible for its own actions."[1]

The distinction between an AI's actions versus its creators' intentions has serious consequences for our court system.

This is not because we're going to throw algorithms in jail, but because humans might deflect blame to them. Some early manifestations of this potential can be seen in facial recognition and discrimination cases. In 2023, the US Federal Trade Commission reached its first settlement involving allegations of biased algorithms, ordering the pharmacy chain Rite Aid to stop using facial recognition technologies that led it to disproportionately call the police on and eject Black shoppers from its stores.[2] If a firm is accused of discriminating against a protected class using an AI-based selection mechanism, courts must determine whether an AI's developers were intentionally discriminating, or whether discrimination was an unintended outcome of the AI's development and deployment.

The legal implications that are emerging from these uses of AIs are profound, and it will take years, and many more cases, to sort them all out. Different legal systems will likely reach different conclusions, but some principles apply broadly. First, AI is created by, used by, and, at least in the near term, directed by humans, so those humans are responsible for an AI's behavior. This is what a Canadian court rightly told Air Canada when it compelled the company to honor its chatbot's too-good-to-be-true offer.

Second, there should be high transparency surrounding the deployment of AI: those impacted by an AI-augmented decision should have the right to be aware of it. A robust, human-operated appeals process should be available to those who may be harmed by an AI-augmented government decision. Like any fair appeals system, it should have reasonable fees and costs, and provide accommodations for those who cannot easily pay.

Third, AI decision-making should be subject to continuous public evaluation and criticism. No AI system should be adopted on an unreviewable or permanent basis, just as public servants are seldom offered life tenure. If the appeals process discovers that mistakes were made, an internal review should be conducted to determine what the AI's role was and how to fix it. Is it a type of mistake that will happen again? Can the AI be improved? Do certain cases need more human review? Should the AI be withdrawn from that task?

Legal rules will be impacted by AI. Many criminal laws are sensitive to intent, requiring determination of both a defendant's actions and their state of mind. Evidence of intent is notoriously hard to gather, and is a much more complicated endeavor than asking a suspect to tell the truth about their state of mind. This standard of evidence is a protective feature of the law, preventing people who accidentally or unknowingly caused harm from unjust punishment. But in the hands of a skilled manipulator, this is a glaring loophole; any defendant who can cast reasonable doubt about their intent can get away with murder, so to speak. Common-law countries and other jurisdictions have spent centuries refining and balancing standards for criminal intent for humans. These will have to be rethought for AI.[3] For example, degrees of homicide are assessed differently depending on the provable level of intent. When a person employs an AI to take an action, intent can become harder to prove because it gives the person another way to argue that they could not have anticipated the outcomes that would result; the AI, they might say, acted on its own.

The adoption of AI in legal and governmental systems shifts at least partial responsibility for decision-making from lawyers and policymakers to AI engineers and technologists. Those latter parties should share in accountability. AI developers (whether public or private) should be incentivized to ensure that the systems they create yield a higher level of accuracy than the status quo. This could be achieved by predicating contracts on a certain level of accuracy, which would be verified on an ongoing basis by monitoring ensuing appeals; that is, an appellate panel should determine how many AI-augmented decisions are overturned following a second, human, evaluation. Finally, to ensure that the deployment of these systems would not result in systemic bias against specific groups, the ongoing reviews should include evaluations of bias. For AI used in democratic contexts, the results of these reviews should be made available to the public.

Someday, we will need to figure out whether an AI can have intent that is separate from the humans that created it. This may sound unthinkable, but many countries have assigned criminal capacity to other superhuman systems, such as corporations. For example, Volkswagen AG pled guilty to conspiracy to defraud the US in its emissions cheating scandal in 2017.[4]

AI could be used to hide unlawful collusion. If everyone delegates their decision-making to the same AI system, then the collusion becomes implicit and much harder to prosecute. This scenario is playing out right now in the US rental market; in 2024, the government took antitrust action against rental-pricing software maker RealPage, charging that its algorithm enabled landlords to collude in increasing rents nationwide.[5] In the future, collusion might not even require human volition. AIs interacting as representatives of human parties might tend towards implicit, illicit collusion while leaving their human directors unaware (or at least with plausible deniability).

Experience leads us to believe that these legal questions will often be resolved by courts in favor of corporations. Companies using AI will often be held blameless, except when their management's reckless indifference or malice is clearly documented (and sometimes, even then). The best way to address this dilemma is to focus on regulating the outcomes, not the inner workings of the system being deployed. Stores that discriminate against customers on the basis of race should be punished, regardless of whether they outsourced their security to people or to AIs. Government agencies that routinely misapply the law should be reformed, regardless of the intentions of any programmer in their technology supply chain. The best way to ensure fair treatment from algorithms may be to create and use more general systems of accountability, judging outcomes without regard to process and without tailoring specifically to AI.

15

LISTENING TO CONSTITUENTS

Elected officials say that they want to hear from their constituents, but they tend to be fibbing—for understandable reasons. Regardless of how committed and earnest a representative may be, it's a lot of work to wade through mass public input. Legislative staff have finite time and attention, and many are already overworked and overwhelmed, so more speech from constituents translates to less listening by legislators. The decay of organizational listening has become a crisis for democracy.[1]

The only recourse for legislative staff has been to make a cursory review of what each person has to say. When a deluge of constituents writes to their representative, those letters are sorted into two piles—one for and another against—then someone measures the height of those piles (or the size of the email folders). That is, unless they're ignored completely.

Public hearings are no better. Those who sacrifice their time to participate in the deliberative process are often afforded just a few minutes to speak. If too many people show up, they (or the legislators) might have to head home before they even get the chance to offer their opinions.

It's plainly impossible for any elected official to sit and listen to the perspective of every constituent on every issue. Ideally, staffers read and synthesize the dozens, hundreds, maybe thousands of incoming communications. They give their boss a useful, concise summary of the most important inputs so they can consider them in the bills they write and the policies for which they advocate. They can pull a few personal stories, powerful anecdotes, and talking points to address the counterarguments they've heard from the public.

These are the kinds of tasks at which modern AI systems already excel. They can digest huge amounts of text—much faster than any human could read—and answer questions about it. They can provide a summary that trims out all the repeated arguments, irrelevant details, and boilerplate language. They can separate individual testimony from form letters and highlight the most unique perspectives. They can cluster opinions into consistent groups, or geographic segments, to help legislators understand pockets of support for different ideas.

AI gives policymakers and their staff powerful ways to interrogate and understand the diverse perspectives of many correspondents; it helps them listen to all their constituents simultaneously. This is more than listening; it's sense-making.

Academics and civil society organizations are actively experimenting with this kind of AI application. US organizations like PopVox and Policy Synth are building AI tools designed to help legislators develop policy solutions with the help of both constituents and experts.[2] In Europe, Make. org has a tool that synthesizes and lets users ask questions about citizens' assemblies.[3] A project with which we are affiliated, the Massachusetts Platform for Legislative Engagement, is using AI to summarize legislative testimony submitted by constituents.[4]

Beginning in 2023, staff in the US House of Representatives have been experimenting with using AI to manage constituent correspondence.[5] In 2024, Indian Prime Minister Narendra Modi said his staff used AI to help synthesize input from millions of constituents while developing his government's twenty-five-year plan.[6] These same systems can also process comments to different government agencies on rule-making processes.

We can imagine this going further. AIs could conduct web and database searches to find out if writers have relevant expertise or an actual stake in the issue at hand. This capacity could help legislators attend to those with the most relevant perspectives, or do the reverse and ignore the perspectives of people without established political power. Constituent profiling raises sensitive privacy concerns, but it's already a reality. It's the type of web searching any staffer would do today before they set up their boss to meet with a stakeholder.

Again, it's not that AI is doing anything novel here. It's just that AI can scale in a way that humans cannot.

AI can also help legislators respond to the concerns they hear from their constituents. A significant portion of legislative offices' time is spent on casework: responding to specific requests for help from individual constituents. This might mean helping an elderly person connect with government services, or assisting people to navigate government bureaucracies. AI can automate much of this process, although it could easily backfire if people only value human assistance.

Of course, a government official will only be responsive if they care what people think. AI could impact this, too, by shifting the incentives to which legislators and other government officials respond. If AI assistive technologies empower more constituencies to pay attention to and weigh in on legislative matters, it could help voters wield greater influence over their representatives.

Citizen acceptance and trust in AI matters as much as AI's capabilities. Will people be less likely to speak if an AI is the one listening? People may want their letters to be read by a human, even if it's just a low-level staffer. People may object if the AI doesn't include their testimony in a general summary. On the other hand, a 2023 study found that constituents expressed more trust in their legislators after receiving AI-assisted responses to their outreach than they did after receiving either human-drafted or boilerplate responses.[7]

AI-facilitated constituent communication might change the role of advocacy groups. Many legislators rely on NGOs (nongovernmental organizations) to aggregate the opinions of constituencies they care about. If those people can easily express their thoughts individually and legislators can easily synthesize them, the role of advocacy groups might diminish. Successful interest groups may focus on educating and activating their members to use AI tools in order to make their voices heard.

The human option will always be there, at least for some of us. Even if an AI is summarizing thousands of opinions, the most committed (and connected) advocates will still have face-to-face meetings through which to elevate their voice. The difference is that AI will afford more people channels through which to reach officials than otherwise would, even if they are only reaching them as part of an AI-created summary.

This use of AI will provide officials with more excuses to disregard the authentic public input they receive. It has always been hard to separate the heartfelt perspectives of individual constituents from those of people who are just filling in the blanks of a prewritten form letter. Generative AI can, admittedly, make it easier for this kind of seemingly personalized testimony to be fabricated automatically and at scale. (As early as 2017, one agency—the US Federal Communications Commission—experienced a flood of machine-generated comments.[8]) Some will raise barriers against inauthentic input, perhaps restricting contact to those who can show up to a meeting or hearing in person or employing schemes for verifying that online comments came from a human.

We caution against solutions that force people to jump through too many hoops to participate. AI should be used to make it easier for all people—not just the rich and powerful—to engage in the legislative process, and prevent elected officials from disregarding the voices of their constituents. Identity verification processes should be minimal, to avoid deterring people who want to participate from doing so. Given that AI can easily summarize incoming testimony, the cost of extraneous input is low relative to the cost of excluding constituents from public engagement.

Implemented effectively, the use of automation in legislative engagement should result in more person-to-person contact. Legislators and staff newly freed up from sorting through emails and crafting boilerplate replies should spend more time meeting with constituents face-to-face in their offices, in town halls, and in deliberative public forums.

In order to walk the line between empowering people to speak out on issues of concern and restraining inauthentic comments and astroturfing, we must watch out for how using AI changes incentives for parties on both sides of a system. Using AI to count up yeas and nays encourages people to use AI to spam comment systems with content skewed to their perspective. AI filters that demote vitriolic language can incentivize more civil and constructive debate. AIs competing for politicians' ears will result in an environment that resembles websites vying for search rank today; they will be prime targets for hacking and manipulation. And, as we will discuss, using AI to find consensus might encourage commenters to cite other advocates and acknowledge sound arguments from all sides.

16

WRITING LAWS

The first known AI-written bill became a law in November 2023, when Brazilian city councilman Ramiro Rosário used generative AI to write a local ordinance about replacing lost or stolen water meters. He submitted it to the Puerto Alegre city council like any other bill, and it was passed without any human edits.[1]

This is how AI should be used in lawmaking. A local lawmaker earnestly seeks to address a significant policy issue, and uses an AI tool to help articulate and draft the policy in legislative language. Other lawmakers debate the draft, edit it if necessary, and vote on the result. In this case, AI is assisting humans by making it easier to prepare a first draft.

Laws have a massive impact on our lives, yet there's nothing sacred or special about their language, other than the need to conform to the legal code and appropriately reference existing laws. As with every other profession, lawmakers will turn to AI to help them draft and revise the text of laws.

Currently, legislators often have no choice but to rely on lobbyists to draft bills, because they don't have the time or expertise to do all the research, writing, and communication work themselves. In any week, they may be asked to vote on proposals on subjects like criminal justice, workplace safety, space exploration, industrial waste management, military procurement, and more. No single person has expertise on such a wide range of issues. But the industries and civil society groups with interests in each of these areas employ lobbyists, who impact what becomes law.

The influence of lobbyists—not just their money but their information and ideas—is so endemic in US politics that political scientists regard lobbying as a subsidy for legislators.[2] Research demonstrates that less professionalized legislatures with fewer staff resources are more likely to copy and paste legislative text proffered directly from outside groups.[3] Many legislators simply can't perform their legislative work—drafting bills, deciding how they will vote on policy—without lobbyist support.

AI will lower the cost, time, and expertise necessary to generate legislation. Legislators can use AI to acquire the sort of specialized information traditionally supplied by lobbyists—which might decrease their dependence on those lobbyists. If we trust AI systems, and trust legislators to use them in advancing their constituents' interests, that could help diffuse power. It could help elected representatives implement policies that faithfully represent their legislative intent, without reliance on self-interested third parties. Alternatively, lobbyists using AI can exert wider influence, take on more clients, or lobby in more than one jurisdiction at the same time.

We'll have to wait and see how effective these tools are, but they are already being built. The software vendor Xcential has been offering legislative drafting products worldwide since 2002,[4] and developed AI-powered legislative comparison tools for the US House of Representatives.[5] The French government has already developed and evaluated an AI for use in parliamentary sessions, and claims that it matches the performance of specialized human legal drafters in summarizing bill amendments.[6] Brazil's federal legislature has used AI to assist in a variety of tasks, including automated routing of tens of thousands of requests for drafting to its Office of Legislative Counselling, and organizing and analyzing proposed amendments to legislation.[7] The United Arab Emirates announced in April 2025 that it plans to use AI aggressively in automatically suggesting changes to its laws.[8]

Applying AI's ability to scale to the generation of legislative text brings many benefits to policymakers. It could let them work on more bills at the same time, add more detail and specificity to each bill, or interpret and incorporate more feedback from constituents and outside groups. In an international context, AI-written law can help global movements propagate policy changes more rapidly across borders. For example, AI could help localize model legislation to many jurisdictions, or every jurisdiction, at once.

When democracies work well, laws are proposed in pieces, rearranged and integrated, massaged and amended, and ultimately horse-traded before coming up for a vote. It can be a challenge to track the authorial intent of a provision across that timeline, be it from a human author or AI. What matters in the end is consistent advocacy at every step of the process, guiding the zig-zagging flow of the legislative process towards a policy goal.

Sometimes, speed matters when writing law. When there is a change of governments, with a new party in power, there is sometimes a rush to make as many policy changes as quickly as possible to conform to the platform of the new regime. AI could help legislators accomplish that kind of wholesale policy shift. The result could be policy that is more responsive to the electorate, or that contributes to more political instability and whiplash.

AI can help make laws clearer and more consistent. With their superhuman attention spans, AIs excel at enforcing syntactic and grammatical rules. They can be effective at drafting text in precise and proper legislative language, or offering detailed feedback to human drafters. Borrowing ideas from software development, where coders use tools to identify common instances of bad programming practices, an AI reviewer can highlight bad law-writing practices.[9] For example, it can detect when phrasing is inconsistent across a long bill. If a bill about insurance repeatedly lists a variety of disaster categories, but omits one category one time, AI can catch that error.

Although this may seem like minutiae, a small ambiguity or mistake in law can have massive consequences. In 2015, the US Affordable Care Act was threatened to be struck down because of a typo in four words, imperiling healthcare services extended to more than seven million Americans.[10]

Syntactic analysis is important, but we should also expect AI to be used to help get the *semantics* of law right: to better match the effect of the law to the intent of the lawmakers, both in broad strokes and in edge cases.

AI can summarize bills, and answer questions about their provisions. It can highlight aspects of a bill that align with, or are contrary to, different political points of view. It can analyze legislation and assess its potential impact. Chile's legislature developed an AI tool to identify conflicts and gaps between proposed legislation and existing regulatory frameworks, pointing out potential problems a new law might introduce.[11] We can even imagine a future where AI can be used to simulate a new law and determine whether it would be effective, how much it would cost, and what its side effects would be. Sweden, for example, developed an AI tool to help predict the national economic effects of policy choices.[12]

This means that AI could help lawmakers understand laws. Some countries legislate with massive omnibus bills that address many issues at once; the US Congress is notorious for bills that are hundreds of pages long. It's impossible for any one person to understand how each of these bills' provisions would work, or their potential impacts. Many legislatures employ analysts who scrutinize and prepare reports on these bills. AI could perform this kind of work at greater speed and scale, so legislators could easily query an AI about how a particular bill would affect their district or areas of concern. Software vendors are already marketing this sort of AI legislative analysis tool.[13]

This means that AIs will likely be good at both creating and finding loopholes in laws.[14] Loopholes arise when the written text of a law translates to surprising real-world effects—at least, surprising to some. Loopholes generally benefit a particular group, and are sometimes surreptitiously planted in legislation for their later benefit. With AI both being used to plant some loopholes and to detect others, we can't predict how this will change the status quo.

Political scientist Amy McKay coined the perfect word for this: "microlegislation."[15] These are the tiny segments of bills, often just one sentence or even less, that produce outcomes narrowly tailored to benefit specific groups, often in ways imperceptible to others (at least at first glance). AI may have its biggest impact on these microscopic changes. Of course, AI will also be able to defend against special-interest-generated microlegislation by discovering the hard-to-detect effects of wordings suggested by legislators aligned with those special interests.

Automated drafting of legislative loopholes requires reform to institutions in a way responsive to AI, but those reforms need not be specific to AI. Forty-three US state constitutions have "single-subject" rules that limit a bill to a single topic, which helps prevent AI (or other actors) from slipping self-interested provisions into legislation.[16] Other legislatures mandate a certain amount of time for legislators to read a bill. For example, the EU Council allows EU national legislatures at least eight weeks to consider new legislation.[17]

One area where we don't anticipate AI making a difference is choosing the right policies: That will remain the domain of human judgment and electoral politics. As much as technocrats and researchers would like to believe in evidence-based policy, the hard choices in lawmaking are much more about values than they are about the kind of analytical synthesis of data and evidence at which computer modeling and predictive AI is well-suited.[18] Machines can help us simulate the likely outcomes of different policy decisions, but someone still has to decide what the objectives of policy should be, what data to collect, what questions to ask of the machines, and how to interpret the result—particularly when the evidence is far from conclusive. Even when machines are asked to fulfill these tasks, for the foreseeable future, humans will steer their outputs.

17

NEGOTIATING LEGISLATION

Many individuals and groups participate in lawmaking: legislative leaders, rank-and-file party members, public-interest organizations, lobbyists, constituents. Ultimately, law is the product of complex negotiations. That involves substantive synthesis of different ideas as well as strategies and tactics of persuasion.

We expect AI strategists to become common, at least in an advisory capacity. This activity is more general than legislative negotiations, of course. An AI agent could automatically walk through stages of policy development, assessment, and legislative strategy, guiding a human legislator at each step or even executing its own recommendations. Suppose that a lobbying concern has enlisted an AI agent to insert favorable provisions into a new bill. In the policy development stage, the AI agent could use generative models to crank out millions of proposed amendments to the bill, each one expressing a different small, but significant, policy change. The AI agent could then use predictive AI tools, or even more traditional econometric models, to assess the market impact of each proposal.

Having picked a handful of proposals with the desired impact, the AI could propose a legislative strategy for passing the amendment. It might recommend the optimal set of committee members to receive campaign contributions and the right interest groups to win over and announce their support, given the specific stakeholders relevant to each policy proposal. If the model determines that those legislative strategies are intractable, it could go back and tweak the original proposals to win more support.

This process of policy development, assessment, and strategy has a multitude of moving parts, many of them unpredictable. AIs of the near future won't be good at autonomously executing lobbying campaigns of any complexity, and they won't have access to those powerful human elites who walk the corridors of power and have influence over legislators' thinking and votes. But this is the process that lobbying firms pursue to promote their clients' interests, and they are unlikely to pass up opportunities to increase the speed, scale, scope, and sophistication of each part of it. Public-interest advocacy organizations and individuals will find assistive uses for AI, too.

Tech companies will launch an arms race to develop AI negotiators that cater to different markets. Services will be tailored to particular industries, negotiating styles, and opponents. Eventually, we expect every legislator to have their own AI negotiation assistants, and for those assistants to start negotiating with each other. AIs will suggest both broad strategies and narrow tactics, and humans using them will need to know their AI assistants' strengths and weaknesses. Just as when humans hire warm-blooded strategists, they will want to select AI assistants that effectively represent their values. This is the classic principal–agent problem.

Time will tell how good AI can become at negotiation and strategy. Like a human, an AI's success will depend largely on its insight into both parties' negotiating position and mindset. This requires it to have access to descriptive information about the context driving each party to come to the table, their responses to proposals in previous negotiations, the information they are receiving about the current negotiation, and so on.

As AI developer Richard Ngo has observed, AIs have superhuman negotiating capabilities.[1] For example, human negotiators need to keep secrets from each other so as not to reveal too much information about their position. Imagine two parties arming their respective AIs with their full negotiating positions, disclosing all their sensitive information to them, then loading them into a secure computer that deletes them both after the negotiation is over. This isn't possible with human negotiators, unless you kill them both at the end. Setting murder aside, AIs can conduct negotiations with ten parties simultaneously, making reference to exactly the same information and starting point of view and sharing information seamlessly across each conversation. Systems and institutions around these AI capabilities have yet to be developed, but they could make negotiation look very different in the future.

Complex negotiations are where there is the most to gain from AI's capability to attend to a multitude of details simultaneously. Legislators engage in multi-agent, multi-issue negotiation all the time, but probably not optimally. Legislative committees debate revisions to bills. Outside groups lobby on amendments. Constituents give testimony. As you might have noticed, many people seem dissatisfied with the outcome of these debates. Part of that is inherent to the opposing interests of different parties, and part is due to the limited bandwidth of negotiators to consider every item that could be on the table. AI isn't going to guarantee widespread satisfaction with the final compromise, but it has a real potential to help each stakeholder get a little more of what they want.

Lastly, there is a concept in statutory and administrative law of "revision." Occasionally the legislature goes through the entirety of its law and cleans it up, removing obsolete and anachronistic provisions, fixing typos and other errors, resolving inconsistencies, and so on. There is precedent for administrative law revision in the US[2] and other countries. When Ireland conducted a large-scale statutory law revision in 2007, it repealed more than three thousand laws, all of which were at least eighty-five years old.[3]

The process of revision is akin to legislative negotiation at light speed. Instead of haggling over a few new provisions in depth, the legislature agrees to make many changes with small impact wholesale. This is generally supposed to be a nonpartisan form of housekeeping, more editorial than anything else. But sometimes revisions result in substantive changes, some accidental and others surreptitiously introduced.

AI can revise laws; it can edit the entirety of a legal code and make it more concise, more consistent, and more understandable. After an AI performs this operation, the revised laws can be voted on by the human legislature. Starting in 2020, Ohio has been using AI to revise its administrative law, trimming 2.2 million outdated or unnecessary words.[4] Assuming an AI can do this well and fairly (or, at least, as well and fairly as a group of humans), it's a good application of AI's speed. The champion of Ohio's project introduced a bill to do the same thing at the federal level in 2025.[5]

Assigning revision to AI means delegating real power to shift policy. Most of the real-world use cases for AI in legal revision to date, including in Ohio as well as Canada and New South Wales, Australia, seem to be focused on achieving deregulation: stripping down the regulatory state.[6] It's not surprising, because there is a strong political constituency for this in many places, and because executive agencies can change their administrative law with less haggling than the voting that statutory law change requires. AI tools could equally be used by leaders seeking to strengthen regulation; for example, to identify cases where regulatory requirements are based on decades-old cost-benefit analyses, and misstate present risks and benefits. The most novel use of AI in revision might be to make revision a more multilateral process. The speed of AI could allow many parties, represented by AI agents, to haggle over thousands of changes to a legal code in parallel.

It's appealing to think that AI could enhance democracy by giving more people, especially those with less wealth and power, access to the ears of legislators. Some of that will happen. But those best positioned to wield AI tools effectively are those with the best understanding of the legislative process: legislators and professional lobbyists. That means that AI will tend to concentrate political influence among the powerful elite unless democracies take steps to inhibit that outcome.

18

WRITING MORE COMPLEX LAWS

Introducing AI to the legislative process could result in more complex laws. This would affect the balance of power between the executive and legislative branches of government.

A common turn of phrase you'll see in many bills is, "the department shall promulgate rules for <blank>." This is shorthand for "the members of the legislature couldn't agree on the details of how to implement and enforce this law," or "the topic was too complex for us generalists," or "we don't want to be blamed by our constituents for those details"—so they left the blanks to be filled in by the executive branch agency in charge of this area. There's nothing wrong with this, but it illustrates an important aspect of the balance of power between these branches.

As legislators gain access to AIs with increasing bill-drafting sophistication, this push and pull between the legislature and the executive could be upended. Legislators can exert more control and articulate more detail about how their policies will be implemented. AI-written laws afford the legislature—and the lobbyists and special interest groups that influence it—greater power to specify detailed policy with arbitrary precision. Human legislators could continue writing broad policy prescriptions, while using AI tools to help address dozens of different special cases.

When the cost of considering and drafting legislative language is essentially zero, the precision of law can increase and the need for rulemaking to fill in the gaps can be minimized. If every legislator has access to detailed technical expertise via an AI, as well as effectively unlimited research capacity, then little discretion needs to be granted to the executive agency. The adoption of AI to automatically increase legislative precision would constrain executive discretion and shift the balance of power between the executive and legislative branches of government.

Legislatures pass laws, and administrative agencies write regulatory rules that interpret and implement them. At the US federal level, there are typically more than three thousand discrete new, final rules published each year. And these detailed rules matter as much as the laws that precede them. Each year, the set of new US federal rules typically includes about fifty to a hundred "major" rules significant enough to have a $100 million economic impact.[1] The EU's regulatory state is arguably even more robust.[2]

In nonparliamentary systems like the US, reliance on agency-drafted regulations matters most in a divided government. When both houses of a bicameral legislature and the president or state governor are controlled by the same party—what is called unified government—then the stakes of leaving rule writing to the executive agency are low for the legislature. Broadly, everyone is on the same page.

But in a divided government, when the legislature and executive branches are controlled by different parties, the significance of rulemaking is much higher and the debate over regulation more intense. Legislators of one party are hesitant to hand control of any significant policy decision over to the executive branch of another party. Empirical studies demonstrate that under a divided government, the US Congress grants less discretion to the president's agencies for rulemaking,[3] and less rulemaking gets done, period.[4] Another study has found that the only factor comparable to divided government in driving complex legislation is partisan polarization; more ideologically extreme legislators introduce more complex bills.[5] These factors often go hand in hand, and can amplify each other.[6]

In the US, the importance of this balance was heightened when the Supreme Court struck down the long-standing "Chevron" doctrine in 2024 that gave deference to executive agencies to interpret ambiguous law. Details left unspecified by the Congress are now the subject of achingly slow court proceedings. Consequently, for the first time in generations, the US Congress does not have the luxury of assuming that the details worked out by the executive agencies will be respected by the courts. The choices before the current Congress are to go back through myriad laws to specify their legislative intent in more detail, perhaps with the aid of AI, or to leave policy up to the interpretation of courts.[7] Given the current dysfunction of Congress, the latter route seems most likely in the near term—but future Congresses will face the same options and may be more productive.

The push and pull between the legislative and executive branches is also evident at the US state level. Over time, governors have assumed more and more power, exercised with particular vigor during the COVID-19 pandemic.[8] And most have a sweeping authority unavailable at the federal level:

the line-item veto. The ability to accept or reject specific sections of policy passed by legislatures, sometimes even specific words, confers truly extraordinary discretion on state governors.

Systemically, it may be a good thing for AI to help legislatures counterbalance the growing power of the executive in the US and elsewhere—especially in countries with authoritarian leaders.[9] Legislatures have the potential to offer citizens more direct representation and greater capacity for deliberation and participation in policymaking. However, this benefit will only be realized if legislatures choose to exercise this power, and if citizens have faith in their legislatures, which many currently do not.[10] And, of course, complex laws written by AI could have deleterious effects; for example, an AI could write legislation that humans can't fully understand without their own AI assistants to explain it to them. Over-reliance on technology could distance humans from the bills they vote on and the laws under which they live.

19

EMPOWERING MACHINES

There are two basic types of laws. Some are explicit, proscriptive rules: "Speed limit 55 miles per hour" or "It is illegal to drive with a blood alcohol level of 0.08% or greater." Others are standards that require interpretation: "It is illegal to drive recklessly."

Rules have long been enforced by machines, but the interpretation of standards is traditionally done by a human. Over the years, court cases involving reckless driving charges have given us a body of definitions of the term "reckless," examples of recklessness, and judicial decisions pertaining to recklessness. If you're charged with reckless driving, a police officer will show up in court and explain what it was that you were doing and why they determined that you were reckless.

Legal interpretation has always been subjective. US Supreme Court Justice Potter Stewart famously acknowledged, in the context of defining obscenity, that sometimes law comes down to "I know it when I see it." That kind of judgment can be administered by an AI as well as by humans.

AI equips lawmakers with powerful new tools for encoding complex standards into law. Rather than delegate authority for this task to other decision-makers in the executive or judicial branches, legislators can directly specify AI systems (or instructions for AI systems) that represent their vision. Laws might even be written in computer language rather than natural human language.

Let's imagine a designated AI taking the place of that police officer. Assume that the jurisdiction has traffic cameras everywhere capturing video footage of millions of drivers, including those who were engaging in observably dangerous behaviors immediately preceding collisions or near-misses: extreme speeding, cutting off other cars, running red lights. We train an AI on this data, teaching it to classify driving as either reckless or not reckless based on the probability that it will lead to harm.[1] Given enough data, the AI could have a more accurate and less biased assessment of reckless driving than human police officers or than notoriously unreliable human eyewitnesses.

Now assume a jurisdiction passes a law that basically says, "Reckless driving is defined by this AI; if it says you were driving recklessly, you were—period." In a sense, the AI is similar to a red-light camera or a breathalyzer: sensor measurements are fed into a computer, which then produces data that can be used to establish guilt or innocence. But in another sense, AI is wildly different. An AI that defines reckless driving could be an opaque black box, its rulings based on millions of parameters, impossible to explain. This is different from the complex verbal laws we discussed in the previous chapter. If the AI can't explain itself, then we humans can't fully understand the law. We would know the generalities, but not the exact boundaries.

This sort of AI-empowered law could arise in many domains if citizens accept it. AI could be used to recognize market manipulation, or medical malpractice, or breaches of environmental law. In some fields, we accept incomprehensible forces, such as pharmaceuticals that work but whose biological mechanisms even experts don't understand.[2] But law is an area where comprehensibility matters, even at the expense of accuracy. The state has a judicial imperative to show cause for punishing its citizens. This is part of due process and exemplifies the principle of *legality*, which requires that criminal law be clearly explained prior to its enforcement. However, the existence of ambiguous and convoluted law demonstrates that this principle is not strictly followed. Moreover, enforcement of law is generally subject to discretion, which makes gray areas inevitable.

Could legality and due process be satisfied without comprehensibility? Could understanding be traded off for predictability? Unlike a police officer's individual judgment call, AI-generated allegations of reckless driving could be consistent and repeatable. That is, the AI could be built to flag the specific driving behavior as reckless every time, whoever engages in it, and the same AI can be used consistently throughout a jurisdiction. This isn't easy: AI systems might produce different results depending on whether there's a bus in the background, or the color of the car in the opposite lane, or the expression on the driver's face . . . but it's possible.

Regardless of whether the AI can explain the decision factors involved in its assessment, we can make its outputs eminently knowable. Perhaps we will embed the AI that determines reckless driving into the cars themselves. Drivers could receive warnings: "Your driving behavior has been classified as reckless." This could be combined with automatic enforcement: "If you continue driving in this manner for another five seconds, you will receive a citation." Or, more ominously: "If you continue for another thirty seconds, your car will be safely deactivated." Or, more helpfully, the AI could justify its assessment with a counterfactual: "To avoid this warning in the future, swerve less." Many cars already give simple automated warnings like this, which could grow more sophisticated with improved AI.

As we discussed earlier, it matters less why the AI reaches the judgment it does and more what impact the AI's judgment has. The consequences of these AI-assigned judgments must be assessed and there should be accountability, and a change in systems, when they produce disparate impacts on specific populations. Drivers from certain countries may be assessed as driving recklessly because driving norms are different in their homelands. People with lower income may be assessed as driving recklessly because they can't afford cars with fancy AI assistance features.

These problems are by no means unique to AI: Humans can be biased and inscrutable, too. But when law enforcement is automated, expect considerable pushback on the legitimate grounds of both bias and privacy, both of which we will talk more about in future chapters.

IV

THE AI-ASSISTED ADMINISTRATION

Once laws are passed, the government implements and enforces them. This is an area where automated algorithms already have a complicated legacy, and it's quickly growing more complex with AI.

The first mandate of democratic government is to serve its constituents. Some of the largest organizations in the world are the bureaucracies of national social programs. Their leaders and the people they serve alike will benefit greatly if their processes are faster, more economical, and more efficient. There are many ways that AI could play a role. Agencies that purchase many billions of dollars' worth of goods and services could become better negotiators using AI. Agencies tasked with enforcing regulations could do so faster and at a larger scale using AI.

Complicating this process is the fact that implementing law involves the application of human values. That is, even when implemented in good faith, law is subject to bias, and punishment could be meted out unfairly. Finding agreement on values, on what is biased, and what is fair is hard enough to do among humans, and will be more difficult when machines are added to the picture. Furthermore, not everyone can trust their government to implement its laws in good faith. In an environment of rising authoritarianism worldwide, many will distrust the intentions of leaders seeking to expand their capacity using AI.

And there's another elephant in the room: Will these AIs replace individuals or will they serve them?

20

BACKGROUND: EXHIBITING VALUES AND BIAS

Any AI will have an embedded value system—and biases—because it is created by humans and trained on human-generated and -curated data.

AI is a tool shaped by humans, and whoever creates an AI system will steer it to reflect their own values. An organization that creates an AI political explainer will want it to reflect that organization's values. A political party will want a campaign strategizer to reflect its values. Politicians using AI to write talking points about their bills will find it unacceptable if the AI is *not* biased to reflect their personal values.

Those are easy cases, but difficulty quickly arises. Think about an AI that is trying to summarize government documents, facilitate a meeting, or administer a government benefit. It's easy to say that we want it to be "neutral," but there's really no such thing. An administrative AI might lend more credence to one side of an argument when summarizing or facilitating a discussion, or it might be more lenient with one party to a dispute than with another.

Ethical values differ across cultures and countries. Several countries, including nondemocratic regimes like China's, have built their own AI models. These models sometimes give radically different answers to the same question, seemingly conforming to their creators' perspective or doctrine. Even within a nation, different political movements can have vastly different ideas of what is a just or unjust policy, not to mention what is a proper use of AI. AIs could be deployed to reinforce a governing party's existing ideology. For example, some AI models built in the People's Republic of China maintain that Taiwan is not independent.[1] Whether this is good or bad depends on your own values.

We humans exhibit bias deliberately (think: paid endorsements) and accidentally (think: eyewitness testimony). We have trouble being objective about other people or even ourselves. AIs may be more or less biased than their human counterparts; they may be biased in different ways, and to different degrees.

Pretty much every form of human bias shows up in AI systems. Current AI can be racist, sexist, ableist, antisemitic.[2] There are many sources of this bias—the training data, the feedback the AI receives during use, the objective functions chosen for the AI to pursue—but the basic source is human nature. Human biases are naturally reflected in what we say and do. Modern AIs train on human activity—often speech—and the human bias becoming embedded in the AI system, whether intentionally or unintentionally. This can apply even in instances where humans are less biased today than we were in the past, or when the AI's designers sincerely want to be unbiased. Because historical data is used to train AIs, past bias can creep into present AI systems.

Patterns of bias leading to discrimination and exploitation can become amplified and entrenched by AIs tasked with examples such as reviewing resumes for hiring, recognizing faces for security, and establishing bond amounts and sentences in courts. Consider the auto insurance market. Insurers using AI in underwriting have been accused of algorithmic bias against minorities, assuming higher risk for those customers and driving up their costs.[3] Moreover, many insurers offer "safe driver" discounts to those who accept automated monitoring systems.[4] This sounds fair, but these "safe" drivers are effectively made to suffer an invasion of privacy to mitigate a baseline rise in the cost of insurance.

There is considerable research on how to recognize bias in, and then debias, AI systems. Results are mixed and work is ongoing, but the commitment of major AI developers to this task is sometimes belied by their actions, such as firing their staff ethicists.[5]

Sometimes bias is deliberately introduced into an AI system by its human creators. Entrenched interests can use AI for "empiricism-washing," dissembling that an answer is more factual or more objective because it was produced by a computer and not a human. Controlling AI systems (and the data used to train them) confers the power to cherry-pick what is presented as objective reality. That is, to position the AI output as what is best for society as a whole, rather than just what is best for those in control.[6]

The ability to easily deploy AIs with different value systems can affect the stability of a government bureaucracy. Today, when a new politician comes into power, they appoint senior government officials who reflect their values. However, a large class of career civil servants generally remains from administration to administration. They tend to slow roll policy changes proposed by elected leaders, sometimes to strictly uphold the law, or to preserve the priorities of the leaders who appointed them, or just to save their own jobs or departments. In Brazil in the 2010s, a series of presidents with diametrically opposed environmental policies each struggled to overcome the inertia of their predecessor's bureaucracy and to embed their own policies in agencies going forward.[7] At the advent of the Third Republic in 1880s France, the newly empowered republicans sought to wrest control of government from the conservatives and Catholics of the day by purging civil servants. Whether you celebrate purges like these as a manifestation of the will of the voters, or condemn them as indices of authoritarianism or corruption, tends to depend on your alignment to the leader making the decisions.

If AI systems begin to replace those mid-level career civil servants, it will become easier for a new government to change direction. It's hard to fire thousands of government employees and replace them with your political allies. Even in places where the law and institutions allow this, mass firings are very visible and generate backlash, and the messy process of replacing those staff leads to disruption of government services. But it's easy to replace a single AI that is doing the work, or even just assisting the work, of thousands. This tension between efficiency over continuity isn't new; what AI changes is how they can be balanced.

In many democracies, AI adoption and automation could amplify the legitimate fears of minority groups that they might be harmed by a new, hostile administration. Whether biased AI systems might be applied in your favor or opposed to your interests, administration by biased AI can be distinguished from administration by biased humans by speed and scale. If government systems are automated by AI, the whole of its force can be redirected to enact new policy much more quickly. This is a real boon to new governments trying to execute their political program after an election, and dangerous in the hands of authoritarian leaders.

It is possible that AI bias can be better understood than human bias. To a greater extent than humans, AIs can be studied. Their biases can be known, and—in theory, at least—corrected (if the AI developers want to, and if they are trustworthy). They can be asked millions of questions and have their answers analyzed. They can be modified, tested for bias, and modified again. We try to do this with human systems as well. In 2020, thousands of pairs of individuals—white and minority—posed as home seekers to test for differential treatment across the US. The study found that whites were consistently and persistently favored by landlords, illegally.[8] These kinds of tests of bias are critical for any fair system, but are onerous and costly to execute with humans. They are far more feasible with AI. Moreover, AIs can continuously monitor for bias in existing human processes.

But AIs can't be completely unbiased—and that's not entirely a bad thing. There are biases we want. A bias for accuracy, for example. Perhaps a bias for kindness. Maybe a bias for fairness, even at the expense of efficiency. Or a bias for finding defendants innocent rather than guilty. Deciding between these biases is a matter of values, making the struggle over the future character of AI a very human-centered one.

21

BACKGROUND: AUGMENTING VERSUS REPLACING PEOPLE

AI will impact tasks generally performed by humans in two very different ways. The first is that AIs can replace people. They can send fundraising emails, summarize constituent feedback, and answer FOIA requests automatically. They will be able to fly planes and drive cars. Perhaps we won't need people for those tasks anymore, because AIs will be capable of doing them.

The other impact is AI's ability to augment people's capabilities and make them more effective. AI can help draft bills, allowing human legislators to propose even more complex legislation. It can augment human responses to polls, making polling faster and hopefully more accurate. It can act as a political consultant, advising candidates and political strategists.

Both of these save human labor, but they do so differently. AIs replacing people means that those people will need to find new jobs. Job loss due to AI will be a major policy problem, albeit beyond the scope of this book. In contrast, assistive AIs' enhancement of human performance could mean people's jobs morph into something more managerial and supervisory—that may also translate into the need for fewer humans to get a job done.

People working together, each with specialized skills and knowledge, make each other better. Research has likewise found that multiple, cooperating AI agents improve the performance of each individual AI. Best of all may be a combination of humans and AI. Civil servants who perform the arduous, repetitive work of reviewing thousands of dull documents will find it useful to have an AI tool instantaneously review them first, then give them a heads-up about what to focus on. Humans may find that useful even if they must supplement or even correct some of the AI's findings.

A couple of questions govern if and where AI enhancement will succeed. The first question is: Who does AI benefit more? Is it the most skilled humans, or those of average ability? The answer depends on the specifics of the task. For AI-assisted call center workers, evidence suggests that AI improves the performance of average workers, but does little to improve the best workers.[1] For computer programmers, some data suggests the opposite. The better the human programmer is, the more their performance is improved with AI assistance.[2]

The second question is: Who is ultimately in charge? Some uses of AI require that a human make final decisions, such as when an office AI assistant drafts an email or a medical AI offers a second opinion on a diagnosis. Giving the doctor final responsibility means that AI mistakes can be caught and corrected, assuming humans are empowered to do so. It can be more problematic to put an AI in charge. Rideshare services and their automated routing systems illustrate the problem: Although a human is driving, an AI is telling them where to go and what to do when they get there. If the AI makes a mistake, like suggesting a wrong turn, the human is likely to blindly follow because they have been habituated to defer to the machine, or are simply not empowered to do otherwise. If your driver follows an AI misdirection, you might be able to live with it. If your radiologist does, you might not.

In the democratic context, it would be great to empower civil servants with access to AI tools that allow humans to outline the policies they wish to enact, then leave the leg-work of implementation to reliable, trustworthy machines. In contrast, it would be nightmarish for AI to dictate policy directives, then rely on humans to interpret what they mean and how they are supposed to work.

We should also be skeptical of delegating the function of public government programs to private AI companies; the privatization of essential public services has a long and con-troversial history.

AI assistance may allow humans to work less, or with greater discretion. The ideal is that humans could choose what to do with the hours they save. They can perform deeper reviews of a thorny issue that they know calls for more research and thought. They can chase down an anom-alous piece of data, which might represent a simple mistake or might be a signal of fraud. Maybe they can review two cases in the time it used to take for one, or spend their time on another task entirely, whether for work, family, or leisure.

This is the potential of an automation dividend. If we can offload more and more of our work to AI, we should all benefit from the time that we, our employers, and our governments save. However, the promise of this dividend is not a guarantee. We will fail if the dividend just translates to higher-volume demands from our bosses. If a social worker with an AI assistant can deliver better care to their patients and have more time off work, that's a win-win. But if that social worker just gets saddled with an even greater caseload, neither party benefits.

Unfortunately, history has taught us time and again that workers tend not to benefit when productivity increases. Often, those gains are captured exclusively by a wealthy few. The worker who produces more each hour is rarely repaid by working fewer hours for the original salary, or being paid more per hour. In the context of government, the result of the AI dividend could be nothing more than higher profit margins for vendors who reduce their internal costs while inflating their public contract bids.

Governments and corporations respond to different incentives. This is reflected in the fact that public sector workers have historically had superior work-life balance compared to their private counterparts, at least across the US and Europe.[3] Corporations have largely obligated themselves to maximize shareholders value, and will optimize themselves to capture all the profits of automation. Governments don't have to do that.

Public agencies can demonstrate how to deploy AI positively, to make their employees' lives and outputs better. But that requires public pressure: We citizens must provide political incentives to adopt AI in ways that support the civil service and make government administration outcomes better, and not just ways that slash jobs and subjugate government employees to AI.

22

SERVING PEOPLE

Democracy is about more than elections and lawmaking. It includes all the bureaucratic ways that citizens interact with the rules of their society. You may not pay much attention to transportation policies while they're under debate, but you will when it comes time to show up at the motor vehicle office and renew your driver's license.

AI has a lot of great potential for enhancing public service, through speeding up, augmenting, and eventually changing many human processes. Although this technology is capable of improving people's lives, it also possesses real hazards. As we'll explore, when we adopt AI to perform a task, our conceptualization of that task and the politics around it may change. If we're not intentional about those changes, they can lead us down damaging policy avenues that concentrate power and exacerbate inequity.

Our first examples are easy wins. First, consider automatic language translation, which we touched upon earlier in the context of politics. AI can help deliver government services in a variety of languages, and can help people get instant support from bureaucracies that might otherwise keep them on hold on the phone, or waiting in line for hours. The city of Gilroy, California, for example, uses AI translation to make its public meetings accessible to non-English speakers.

Next, consider traffic. Managing roads and traffic is a major responsibility of local governments, and a good use case for AI. Google's Green Light project has applied AI to traffic signal optimization in cities across four continents, reducing stopping times at intersections as well as emissions from idling vehicles. Boston has reported over 50% reduction in traffic at two busy intersections using this tool.[1]

Applications like benefits administration are far more complicated than traffic signals. They involve creating criteria for receiving a benefit—whether all-or-nothing or a sliding scale—then deciding who is eligible and who is not. It's a laborious process, both for the citizen who is trying to prove eligibility and for the civil servant making the eligibility decision. It's also prone to error.

Computers already use simple rules to determine eligibility. AI could streamline the process further, working with messier data and more nuanced criteria. It could help applicants complete forms, instantly assessing how they will be reviewed, warning users about errors or responses that indicate ineligibility.

There are many places where we don't hire enough available humans to process claims. In the US, an alarming example is the Social Security disability benefit, where the average wait time for an initial decision on a claim is nearly eight months; in 2023, 30,000 people died while waiting for a decision.[2] (The number for the equivalent program in the UK is lower, but still tragic: 1,700 per year.)[3] So long as there is an opportunity to revert to manual claims review through an appeal, using AI here should be a win. We can expect AI to be faster than the existing, achingly slow system, but we cannot expect it to be more fair. Manual claims review is itself often biased, arbitrary, and unjust, and automation itself does not address that. Political scientist Virginia Eubanks has coined a term for this: "automating inequity."[4]

Speeding up bureaucratic decisions with automation is not as easy as building an AI and flipping a switch. It is notoriously hard to adopt new technologies in government. As the former US Deputy Chief Technology Officer Jennifer Pahlka has explained, there is often a wide gulf between policy and implementation.[5] AI is not immune to that problem. In some cases, AI will displace older technologies already overdue for modernization. Sometimes, AI will be thrust onto agencies whether they want it or not. Microsoft, for example, has been pushing AI onto government licensees of its Office cloud suite since 2024[6] and OpenAI released a version of ChatGPT for governments in 2025.[7] Established vendors, especially Big Tech companies, benefit because they can add AI into products already in use; upstart offerings may need to complete arduous and protracted security and procurement processes.

We expect the natural bureaucratic early adopters of AI to be smaller, more nimble, more digital countries such as Estonia and Taiwan. Meanwhile, the enormous bureaucracy of the US shifted ideologically—and recklessly—towards AI at the start of the second Trump administration, initiated by the hostile takeover of federal agencies driven by Elon Musk.

Beyond the software that public agencies use internally, governments around the world rely heavily on contractors to implement and, often, to operate systems on their behalf. These vendors seem primed to rapidly adopt—and sell—AI integrations. As a result, government procurement rules will be the most immediate form of AI policy that many governments manage. Regardless, for most governments the bureaucratic transformations we describe in this chapter will be years-long, perhaps decades-long, projects.

Like any system, AI can be implemented poorly and cause harm. More than half a million Australians were incorrectly ordered to pay back welfare benefits based on a faulty algorithm from 2016 to 2019.[8] In the US, the consulting firm Deloitte won hundreds of millions of dollars in contracts to build automated eligibility determination systems for state-administered health insurance (Medicaid) before the judge in a class action lawsuit ruled that their system had improperly denied coverage to thousands.[9]

Experiences like these, and numerous examples of AI exhibiting bias and encoding racial discrimination, have led scholars like Arvind Narayanan and Sayash Kapoor to recommend a blanket prohibition of the use of AI to predict social outcomes, such as criminal recidivism or the probability of loan repayment.[10] They argue that predictive optimization tools lack legitimacy: they make more mistakes than they claim and they transfer decision-making authority from elected representatives to technocrats.[11] However, this conclusion discounts the reality that we already use automated systems to predict social outcomes in decision-making processes. Traditional human systems, often institutions with inscrutable policy rulebooks and vast bureaucracies, suffer from similar shortcomings of legitimacy.

Our recommendation is more nuanced: to always consider the social impacts of empowering AI decision-makers. Every application of AI trades off potential risks and benefits with the human-operated system it would supplant, and many applications of AI that exacerbate discrimination and harm could instead be used beneficially if applied purposefully, carefully, and with accountability. AI systems performing tasks of social consequence should not be evaluated as static objects in isolation, but in the context of alternatives to preexisting systems and alternate configurations of the same AI tools.

Regardless of what a government does, this kind of automation will occur in the private sector. Corporations will use AI to make all sorts of decisions: Should insurance pay a claim? Should a loan be approved? Should a refund be issued? We would like government services to be equally efficient and responsive. Even if they're not, people will use AIs to help themselves navigate government and commercial systems. Consider Holden Karau, an engineer in San Francisco, who built a website using AI to help people write appeals of health insurance claim denials.[12]

As AIs become better and more accurate, policymakers could harness algorithms to consider ever-more-complicated eligibility criteria. Future lawmakers (or civil servants) would be able to create even more sophisticated rules and to target benefits even more precisely. An AI-powered benefits administrator could base its decision on hundreds of factors, and take into consideration any number of extenuating circumstances—something human administrators have trouble doing.

This capability to precisely target benefits will be attractive to some, particularly those looking to trim social spending. For others, it would be dystopian. Already, nutritional benefits in the US under the SNAP program can be used for fruits and vegetables, but not hot or prepared foods. Already, AI is widely used by US states to determine SNAP eligibility and detect suspected fraud, often with poor implementation that causes substantial harm.[13] Imagine AI systems that use health metric data to let you buy ice cream only if you've consented to wear a fitness tracker that reports that you've exercised enough that week. Political incentives exist for restrictions like these, abhorrent though they may be. Legislators will want to advantage their constituencies, perhaps by denying benefits to people in other voting blocs. Interest groups will want AI to reflect their values, like restricting sugary foods. Industry lobbyists will want their products exempted from restrictions; maybe sugary yogurts will be deemed acceptable. All these pernicious incentives predate AI, but will pervade its use. New technology gives policymakers greater speed, scale, scope, and sophistication for implementing ever-more-proscriptive rules.

Our next example is public education. All of the conversational capabilities we talked about earlier for politics also apply to government communicating with its citizens.

Governments have already begun to use AI tools for educational purposes, with the first applications being to streamline service delivery. The demand has been there for a long time. By 2020, perhaps driven by the COVID-19 pandemic, the developed nations of the Organisation for Economic Co-operation and Development reached an interesting milestone: The majority of them had implemented virtual assistants, AI chatbots, to respond to taxpayer inquiries.[14] There are now innumerable examples of AI being used in this way. Many governments, like those of the UK,[15] Canada,[16] and the state of New Jersey,[17] use AI chatbots to explain some government registration and regulatory processes.

Using AI in sensitive contexts like government service delivery and public education raises significant ethical concerns, particularly with respect to factual accuracy. For example, we don't want AI chatbots that give people incorrect instructions about how to vote. Even if they improve over time, it will remain a significant problem if AI confidently provides incorrect information, even if only occasionally. The difficulty, of course, is that people have increasingly divergent ideas of what is fact or fiction, particularly around sensitive topics, and AI can't help with that.

All of these shortcomings are real, problematic, and apply to humans as well. The people upon whom governments rely to deliver services and inform the public also sometimes misrepresent the facts, sometimes accidentally and sometimes purposefully. Even so, we have useful government agencies and effective services because we are able to build systems that are resilient to these individual points of failure. Similar approaches can be taken to integrate AI into effective organizations. Outputs from AI systems should be monitored to comply with standards that are at least equivalent to those required of humans in the same decision-making roles, with redundancy and error-checking systems incorporated to catch mistakes when they happen.

For a lifesaving use case, consider disaster management. Since at least 2021,[18] the US Federal Emergency Management Agency has used computer vision tools to comprehensively assess damage after hurricanes, tornadoes, and other disasters.[19] This information is used to target support resources and unlock relief funding. After Hurricane Ian devastated the southeastern states in 2022, the agency says that this use of AI reduced the number of structural assessments requiring human review from over a million to fewer than 80,000.[20]

The efficiency of patent offices could also be greatly enhanced by the use of AI. One of the most time-consuming jobs in patent review is the search for prior art: previous inventions that can invalidate a patent. AI can bring both speed and sophistication to this task and help keep pace with demand by enabling examiners to move beyond keyword searches to more robust queries based on a patent application's semantic content. Brazil's patent office claimed to reduce its patent examination backlog by 80% by using an AI-assisted workflow.[21] The World Intellectual Property Organization has tracked initiatives to use AI in similar ways in at least twenty-seven countries.[22]

AI can also help governments fulfill freedom of information requests faster and with fewer resources. Freedom of information laws are pillars of government transparency around the world; they allow individuals and organizations to request a broad range of records, such as agency staff emails or regulatory enforcement logs. The scope of these requests can be vast; the US federal government fields about a million FOIA requests per year, each of which can yield thousands of pages of documents.[23] It takes tremendous effort to sift through agency systems and records to locate material responsive to an information request, to review that material for relevance, and to properly redact personal and classified information that could cause real harm if overlooked. If trustworthy AI tools are available, it will be increasingly implausible for government agencies to charge exorbitant fees, to delay fulfilling requests for months, or to reject requests as infeasible. The US State Department says that it reduced staff hours in making document declassification decisions by 60% by adopting AI.[24]

These examples are far from exhaustive. Any government is filled with humans performing tasks involving research, analysis, review, classification, and summarization—all activities that can benefit from AI.

Despite the fantasies of some, we don't anticipate that AIs will replace the humans who perform these tasks anytime soon. Nonetheless, over time, we expect that AI will make civil servants more effective at their jobs, and democracy more responsive to its constituents. Administrators and policymakers need to ensure that these efficiencies make government serve people better and more equitably.

23

OPERATING GOVERNMENT

Governments around the world, from the local to national level, are eagerly evaluating what processes and tasks they can infuse with AI. By 2020, about half of US federal agencies had already begun to implement AI automation. By 2022, there were at least 250 documented governmental applications of AI in the EU.[1] By 2023, the US had disclosed 710 AI use cases across its agencies,[2] and by a year later, that number had tripled.[3] A 2024 survey by the government of New South Wales, Australia, found 275 implementations of AI for official business.[4] Suffice it to say, governments are not wanting for inspiration.

With thousands of documented use cases in planning or implementation around the world, it's hard to know where to start in describing how AI will impact the operation of government. We will outline use cases that exemplify each of the key factors where we expect AI to make a difference: speed, scale, scope, and sophistication.

Let's start with activities that governments are perfectly capable of doing, but would benefit from doing much faster. Scanning, sorting, and interpreting paper documents at high speed is critical to postal services, for example, which motivated many early advancements in computer vision. Hard-copy archives are being digitized to help automate processing, but slowly. As of 2023, only 2% of US government forms were digitized[5] and many governments have centuries of paper backlogs they must occasionally sift through.

Modern AI tools are being put to use to quickly process decades of old documents. In 2021, California passed a law eliminating racial covenants, which are deeded restrictions on property ownership and sale. The new law required Santa Clara County to review and revise 84 million pages of property records. Two clerks were available to do the work; the county estimated that it would take them twenty years. A research team from Stanford built an AI model that finished the job in six days, flagging unlawful language in deeds for human review and action by the county. Performance testing found that the AI system caught 99.4% of all racial covenants and did not wrongly flag any lawful language.[6]

More generally, Singapore's Open Government Products team has used AI to automate searching government databases. Their tools serve 20,000 active users per week and—they claim—save nearly half the time previously spent on these tasks.[7]

Next, consider auditing for fraud. Human auditors, at government agencies or elsewhere, typically review contracts or accounts only once a year at most, and usually sample just a few of the many contracts or accounts they oversee. Much of the fraud that is detected is caught passively, such as via tips from whistleblowers. A study of fraud across 133 countries found that the best way to decrease losses from fraud is through continuous automated transaction monitoring, which enables fraud to be detected before someone just happens to notice it and chooses to report it.[8] AI auditing, if it is competent, can do better than human auditors because of its greater speed. It can regularly audit *all* contracts and transactions, flagging anything suspicious for review by human auditors. A 2020 demonstration by researchers using Australian government agency records achieved 96% accuracy in detecting noncompliance with procurement procedures, while processing records about ten thousand times faster than human auditors.[9] In 2024, the US Treasury claimed that it had saved $4 billion in one year by using AI to detect fraudulent and improper payments by US agencies.[10]

Let's next consider scale. Governments often serve millions of people across huge territories, so scaling logistics—managing the flow of physical goods—is critical, and right in the wheelhouse of AI. The US military's first operational deployment of AI was for logistics, the Dynamic Analysis and Replanning Tool. DART allowed the military to automate the generation of complex logistical plans for moving personnel and cargo across continents. A US official claimed that DART was so crucial to the Desert Shield and Desert Storm operations in the Middle East in the 1990s that it paid back thirty years of Department of Defense investment in AI.[11] On the civil side, Canada's National Research Council has established an AI for Logistics program that has funded optimization of road and rail networks, developed resilience to transport through extreme weather, and supported other performance-enhancing applications of AI to logistics.[12] We shouldn't take every rosy assessment of AI's success in this area at face value, but logistics has become so closely connected with AI research that we can be confident of aggressive application of AI for years to come.

Tax collection is another area where the scalability of AI will matter, both to assist taxpayers to help eliminate noncompliance due to ignorance (which we already discussed) and to audit taxpayers to help eliminate noncompliance due to cheating. Many governments are enthusiastic about this use case, particularly developing nations and nations that have long struggled with evasion.[13] Italy[14] and Türkiye[15] are among the countries to publicly announce that they will use AI to crack down on tax evasion and fraud. The results are not yet in on how much this is helping, and we cannot expect AI to be a panacea to all the policy challenges of tax collection, but this should be a natural place to test return on investment because the revenue implications are so direct. Historically, the number of tax agents in a country has been a strong predictor of the robustness of its collection.[16] However, even if AI assistance can make those tax collectors more efficient, it's not the tech that matters most here; it's the politics. Hiring tax collectors is a function of political will, as is acting diligently upon noncompliance that might be uncovered through automation.

Third is scope—AI's ability to do many different things—which brings us to the example of procurement. Any large government has an enormous number of procurement contracts with suppliers of everything from police cars to paperclips. The US federal government spends two trillion dollars on procurement annually.[17] This requires negotiation at a massive scope, haggling with many vendors for an endless array of goods and services.

Contract negotiation involves both coming up with options and anticipating how the other side will respond. Because of the subject-matter expertise and familiarity required, human negotiators tend to specialize in one distinct area: maybe real estate, or medical products, or machine tools. AI systems can be trained on millions of human conversations and reactions and tuned to anticipate human response patterns to the text they generate, across any subject matter. Large-scale experiments have already been conducted with AIs negotiating both with other AIs[18] and with humans.[19] As early as 2022, AIs have been playing the seven-player game of Diplomacy, a classic test of human negotiation. AI players have succeeded in convincing human players that they are human and regularly beating them.[20]

We don't expect that AIs will take over all human procurement and other negotiations anytime soon, but they could certainly take over some and assist humans with others. There are assistive and automating roles for AI across the whole procurement task chain.[21] AIs could write requests for quotes, benchmark prices, model costs, define contract terms, and so on. We don't know which of these tasks AI will be most effective for, but the scope AI offers here is enticing. For a start, AI can assist human negotiators by arming them with better information. Since 2019, Peru has used an AI-augmented Electronic Quoter to help its two thousand public agencies estimate prices for procurement.[22] A human expert might have expertise in one purchasing category, and work at one agency. A skilled AI could conceivably help every category at every agency.

Finally, we come to sophistication—AIs taking more factors into consideration than a human could. This capability, too, might lead governments to adopt AIs in negotiation. In the future, AIs might develop durable advantages over humans in strategic negotiation. A fundamental human limitation is our ability to keep only a handful of variables in mind at once, so we tend to focus on a few key issues and tactics. AIs' far greater sophistication has helped them excel in abstract strategy games like chess and Go. Astrophysicist Bill Press credited a machine for discovering a novel, dominant strategy for winning the Iterated Prisoner's Dilemma, a simple game that has long served as a testbed of negotiating tactics.[23]

Even more promising, AI systems can improve their negotiating tactics over time, without human assistance. Just as an AI learned to beat world masters in Go by playing against itself before it encountered human opponents, research has demonstrated that AI models can learn effective negotiating strategies through self-play.[24] They can even become useful in teaching humans how to negotiate.[25] Like skilled human negotiators, AI agents can incorporate theory and analytical strategies that have been demonstrated to make a positive impact on the process.[26] Evidence suggests they can even choose persuasive tactics based on predictions of humans' emotional states.[27]

AIs that negotiate on behalf of governments may represent a mechanism for concentrating power, or even a vector for corruption or hacking. One study found that some AI models are susceptible to offering concessions when their opponent pretends to be desperate.[28] If governments rely on commercially built AIs to engage in negotiations, those companies might surreptitiously introduce intentional failure modes that private parties can exploit to gain advantages. When AI negotiators do not work well, it will benefit those parties who can afford better, and human, negotiators. If they do work well, it may still benefit those entities able to afford a combination of the best humans and the best AI. Even if that advantage is small, it may accrue to corporations—who may be able to more rapidly adopt AI negotiating tools than slower-moving government agencies—at the expense of the public.

Making changes to government operations should be slow and deliberate. Incorporating AI into bureaucratic tasks is not a simple matter. As we have already discussed, there's a big difference between envisioning and realizing a technological improvement to a government process. All these integrations of AI will come in fits and starts, likely with numerous failures along the way. The greatest failures will come when AI is applied suddenly, recklessly, and without change management, public consultation, or human review of data inputs or decision outputs.

24

ENFORCING THE LAW

There has been widespread, understandable, resistance to
the use of AI in law enforcement, because of legitimate concerns about biases and the ethics of handing over decisions
about individual liberty to a machine. Yet automated systems already assist the functions of law enforcement: think
speed trap cameras and breathalyzers that gather the data
that will help determine your guilt or innocence.

AI enforcement can be much more sophisticated than
these task-specific devices. It can automatically identify people who cheat on tax returns or on applications for government services. It can watch all the traffic cameras in a city and
automatically issue citations, even for violations that involve
more complex judgment than simply exceeding the speed
limit. It can look for suspicious investment patterns indicating
insider trading. On the one hand, this use of AI would require
digital surveillance on a massive scale and would trigger widespread public concerns about privacy and totalitarianism. For
example, imagine a regime using AI to trawl the social media
accounts of residents for the purpose of detaining or deport-
ing those who expressed opposing political views, or partici-
pated in protests. This is the kind of totalitarian technological

social control that China has long been accused of developing,[1] and it was reported that the second Trump administration attempted this in the first months of 2025.[2]

On the other hand, this is just AI performing the tasks of existing human institutions, and automating the inequities those humans have decided to exact at greater speed and scale. As we said in the introduction, government itself is a superhuman machine for regulating individual human lives and actions. It's not necessarily more invasive or repressive to have AI monitoring every closed-circuit camera in a city than to have people doing the monitoring, except that we generally don't hire enough people to devote that much attention to the task.

Details matter here, as does system design. Is the AI making human law enforcement staff more effective, or replacing them? How are the inevitable mistakes of AI handled? Whose perspective does the AI represent and reinforce?

In some areas, assigning AI responsibility for monitoring legal compliance works well. South Korea employs extensive automation of traffic enforcement—as of 2023, there were more than 2,000 automated enforcement cameras in Seoul. The Seoul Metropolitan Police Agency has reported an 18% decline in traffic accidents at locations where the cameras were installed.[3] The lack of human traffic enforcement is striking to many visitors.

In other areas, automation in law enforcement has been a disaster. AI false positives can be hard to contest if the courts believe that the computer is always right. The British have experienced this failure rather spectacularly over the past few decades. Horizon is the accounting and inventory software developed by Fujitsu and adopted by the UK Post

Office in 1999. Over the following sixteen years, software flaws led to the unjust prosecution of nearly 1,000 postal workers on charges of fraud, theft, and wrongful accounting. Many went to prison.[4] This failure was not the result of AI—more traditional software was to blame—but it illustrates the harm created when the output of a flawed, untrustworthy system is taken as gospel.

It can be hard to contest an authoritative accusation. Trusting human police officers and prosecutors can be similarly problematic. A primary problem, whether an accusation is made by a human or a machine, is that we often don't know how they arrived at their decisions. The law has always been subject to the whims of its interpreters, and black box AI models exacerbate this.

Bias is another problem, for which both humans and AI are notorious. The difference with AI is one of scale. A biased police officer only interacts with so many people. A biased AI can affect everyone, all the time, like a systematically biased police precinct.

The societal imperative is to build robust systems for human decision-makers, even though these systems come with complex, variable, and sometimes inscrutable biases. In general, people strongly prefer human decision-makers on questions of moral significance.[5] We've developed safeguards for humans, with varying levels of success. We use double-blind lineups to limit the impact of unreliable witnesses. We have rules of evidence that establish the admissibility of information in court, and are aimed at preventing irrelevant or unsubstantiated facts from biasing judges and juries. And we have an appeals process so that wrongful outcomes can be revisited.

Many of these processes could similarly apply to AI, and we should use the same type of systems to constrain its biases. Clear rules of evidence are important, whether that evidence is brought forward by an obviously fallible, human prosecutor or police officer or a seemingly infallible and objective AI system.

If an AI system places a suspect at the scene of a crime through analysis of security camera footage, we should not take the allegation at face value. Institutions should establish the provenance of the video footage, whether the camera or image could have been manipulated, whether the facial identification was reliable and unbiased, and whether the AI could have simply made a mistake. We should treat AI systems as unreliable narrators, scrutinize them as we do human witnesses, independently reviewing the evidence underlying any allegation.

Where adopted, the scale of AI law enforcement has the potential to change our relationship with the law. If implemented oppressively, it could reduce our respect for the law. If implemented unreliably, it could reduce our trust in the law. And if implemented unevenly, it could change the deterrent effect of law.

To be effective, a punishment needs to consider both the damage done and the probability of being caught. For example, if the penalty for smuggling a $100 item is $1,000, but there is only a 5% chance of getting caught, then the crime is profitable in the long term. This is why most drivers in many countries speed: There's a small penalty if caught and the chance of being stopped is very low. If AIs monitor traffic cameras and issue citations every time someone violates a traffic law, then drivers' habits will change dramatically. In

Norway, drivers are conditioned to anticipate enforcement, and compliance with the speed limit near traffic cameras is greater than 99.999%, even if the cameras are turned off.[6]

If AI is given primary responsibility for law enforcement, many laws would need to be revised to take changed deterrence into account. Fines could be reduced in consideration of the fact that lawbreakers will always be apprehended. However, that turns fines into fees that the rich can easily afford but the poor cannot, unless the fines are indexed by income or wealth, or the punishment escalates with repeated infractions.

People will likely rebel against automatic enforcement. In 2021, South Korean privacy advocates succeeded in shooting down a proposal to use facial recognition AI with CCTV cameras for COVID-19 contact tracing.[7] In some cases, lawmakers will choose to protect the poor, or other vulnerable communities, by preventing automated enforcement. In 2024, the New York state legislature banned New York City's transit agency from using AI facial recognition to enforce fare evasion on the city's subway.[8]

Political will is another limitation to automated enforcement. When there is already a cultural norm to violate a rule, being suddenly held accountable for it can feel unfair. In our home state of Massachusetts, it is illegal for the police to use either red light or speed cameras to issue citations. In France, where speed cameras were rolled out aggressively a decade ago, opposition to this enforcement is fierce. Over those ten years, nearly 50,000 speed cameras were vandalized, destroyed, or deactivated by people resisting enforcement.[9] In the US, it has been a challenge to fully fund the IRS because of political opposition, even though catching tax cheats is one of the most profitable things the US government can do.[10]

In some areas, this kind of automated law enforcement will encourage worse behavior. Imagine the example we discussed in chapter 19, "Empowering Machines": automatic detection of reckless driving, and cars that warn drivers when they are driving close to the line. That kind of constant feedback could empower drivers to drive in a manner that just skirts the definition of recklessness, driving less carefully than they would without the warnings. Alternatively, the AIs might obey the law even when humans around them don't. In some early tests, driverless cars would follow the speed limit even when it was an unsafe choice given the speeds of other cars.[11]

We see this kind of behavior by trolls who have learned to post material on social media that is just at the edge, but never triggers the platforms' algorithms for taking down posts. Their behavior is still trollish, but subtle in the exact ways that bypass the filters. It's a perverse outcome for systems designed to make social media sites less toxic. (In China and elsewhere, people game censorship algorithms in this manner.)

Finally, as we discussed earlier, future laws might become so complicated that only AIs would be able to decide whether they are being broken. Already, many companies, individuals, and even government agencies rely on legal teams of lawyers to advise them whether their potential actions would break the law. Interpretation of laws by AI would give more people access to that sort of detailed guidance in advance.

We don't view machine-enforced law as a scary future scenario, but rather as a mundane, if frequently pernicious, fact

of life. The question is not whether people should be sub-
jected to inscrutable, faceless legal decision-making—that's
the status quo. Rather, the question is how to improve policy, using new technologies when helpful, to incrementally
improve our existing systems of justice. AI can afford advocacy organizations and other private actors the opportunity
to help hold government accountable for delivering justice.
We will discuss this more in a later chapter.

25

ENFORCING REGULATIONS

Democracies have no shortage of regulations, but often lack the time, resources, and willpower to enforce them. The result is that wealthy and privileged individuals and organizations ignore regulations with impunity. AI can change this by decoupling the capacity to enforce the rules from the financial resources historically necessary to do so.

Regulations are the rules that the executive branch makes up to implement the requirements passed into law by legislatures. We talked in chapter 18, "Writing More Complex Laws," about how the three branches of government often engage in power struggles over who gets to make which decisions in this power-sharing framework. In general, the legislative branch defines the broad legal framework, and executive agencies fill in and implement the details.

The incentive to comply with regulations depends on the likelihood of detection, the rate of enforcement, and the penalty for noncompliance.[1] While AI automation does not itself change penalties, it can boost rates of detection and enforcement without the need for more staff. Even an agency with few staff and a shoestring budget can, with AI automation, apply its rules and policy judgments to an effectively unlimited number of regulated entities. Regulatory enforcement could become more uniform and thorough, without regard to any ebb and flow in agency budgets. This would make enforcement more scalable and efficient, with significant implications for fairness and public policy. It could also result in an enormous shift in the balance of power between the government and corporations.

This is enticing to us because, in our own research, we've witnessed the weakening of enforcement that follows agency staff cuts. Our analysis of Massachusetts's environmental regulator found that its volume of enforcement actions tightly tracked its budget, with an 80% correlation, as its funding level swung by 70% from 2004 to 2017.[2] This illustrates that the laws passed by legislatures can have little impact without significant resources to enforce them.

In principle, AI regulatory enforcement would mirror human enforcement. A process would generate data on the regulated activity, such as physical inspections or reports that regulated entities are required to submit. Another process would review that data and determine whether it complies with applicable rules, permits, and variances. Finally, when violations are detected, a third process would decide what enforcement actions or penalties should be levied. But AI regulatory enforcement would be different across the dimensions of speed, scale, scope, and sophistication.

As in law enforcement, automated regulatory enforcement is not particularly new. Environmental regulation has been an early beneficiary. The World Wildlife Fund has applied AI to satellite radar imaging to detect and prevent illegal logging,[3] Governments around the world use automated sensors to monitor—and enforce—water quality standards,[4] many of which already integrate machine learning capabilities.[5] Researchers have demonstrated the ability of AI-powered computer vision to identify manure dumping that violates factory farming regulations.[6]

Another long-standing application for AI is anomaly detection. This could be used to detect unusual patterns of behavior by regulated entities that could indicate fraud or the exploitation of regulatory loopholes, then to report those patterns to human regulators for further study.[7] Think of it as an early warning system for bad behavior.

AI can do even more. Although it can't show up for an unannounced inspection on a factory floor, it can inform the movements and data collection of human inspectors, by prioritizing a list of worksites for human inspectors to check. It can review sensor data like camera footage and meter readouts to detect patterns of compliance, disruption, and violation. It can even apply discretionary policies, to promote the political preferences of an administration, to manage a budget for enforcement actions, or to balance the costs and benefits of an enforcement intervention.

Some research suggests that AI can't yet perform as well as humans. In a 2024 experiment, Australia's financial regulator found that no generative AI tool could analyze auditing documents nearly as skillfully as human analysts.[8] We can expect AI tools to improve, but even if they do, the use of AI for automating enforcement activity is not necessarily any more objective than traditional measures. It still involves political judgment; an AI system deciding whether to penalize a violator will still need to make that decision according to a policy outlining when intervention is worthwhile and when it is not. This process would still involve imperfect data, and violations may be missed when the signals collected don't show the complete picture.

Corporations will also benefit from those same capabilities. If this means that companies become better and more efficient at complying with regulations, then everybody wins. But they may use AI to skirt those rules. Probably it will go back and forth over time, with companies sometimes evading regulations and sometimes getting caught.

We can anticipate several macro effects associated with AI-enhanced regulatory enforcement.

First, automating compliance will make it easier for businesses to ignore the spirit of the law. The $30 billion compliance management industry already helps businesses adhere to government regulations across jurisdictions.[9] Those companies are integrating AI to increase the scope of auditing, adapt to changes in regulations faster, and assess risks with greater sophistication.[10] This will change the nature of compliance. When humans are directly involved with compliance certification, they must be intimately familiar with the rules and the consequence of every decision. If they can outsource certification to an efficient machine, they don't have to think about it at all. The system can figure it out for them, and that system won't care about the spirit of the law. This may lead to the discovery and exploitation of new loopholes, benefiting corporations at the expense of the rest of us.

Compliance automation will benefit some enterprises more than others. It could make it easier for small companies to comply with regulatory frameworks. Large companies sometimes lobby for regulatory barriers purely to burden upstart competitors; AI support could help reduce those barriers.

Second, if AI systems are used to automate the interpretation and application of regulations, those regulations could become more complex. A human regulator may be limited in the number of provisions and requirements they can check for compliance, but the scale and scope of an AI review could be essentially unlimited. Human regulators typically prioritize their attention across regulated entities based on either a risk or impact assessment, or random sampling. AI reviews could apply equal scrutiny to unlimited numbers of organizations and rules. And just as for legislation, AI could assist in drafting complex regulations. It could suggest new provisions and assess their likely impact, and the feasibility of compliance. An AI simulator could even anticipate potential ways a regulated entity might exploit draft text, and suggest revisions accordingly.

Third, the same tools that governments use to help review regulated industries could be made available to the public and the media, making it easier to provide public oversight—and impose greater reputational risks—to firms that break the rules. There are already numerous NGOs and entrepreneurs engaged in corporate oversight, sometimes incentivized by government. The US Securities and Exchange Commission pays out billions of dollars in bounties to whistleblowers who report financial crimes,[11] creating a market of fiscal watchdogs.[12] Automating this private enforcement with AI could equip companies with new tools to undermine their competitors.

Finally, regulated entities will still seek to prevail over their regulators through lobbying and the courts, no matter how AI is used. It doesn't matter who is better at enforcement or evasion if one party can rewrite the rulebook.

Expect pushback on AI-enhanced regulation by those who are already powerful. Just as agricultural companies lobbied for laws that prohibit activists from secretly filming the operations of factory farms, we should expect similar lobbying against AI-enhanced regulatory enforcement.

Regulations can penalize some groups more heavily than others, and that bias can be automated with AI. Governments strategically shirk their enforcement responsibility in order to economically advantage some and disadvantage others.[13] Often, this kind of selective enforcement is associated with politics, or even corruption. There are copious examples of politically connected firms receiving favorable treatment, such as being awarded government grants or receiving protection against new entrants in their field.[14] Those preferences could be encoded in instructions to an AI.

AI-assisted lobbying of agencies could undermine regulation in the same ways that humans do. This argues for ensuring equal access to AI tools. If regulations become so complex and burdensome that only the largest corporations can afford the AI tools necessary for compliance, then that complexity becomes a barrier to entry.

Perhaps more so than any other topic in this book, the impact of AI on regulatory enforcement will vary drastically across countries and cultures. Nations skeptical of corporate power will embrace more exhaustive and rigorous enforcement of regulation using AI, and nations skeptical of government oversight will reject it.

Within nations, the automation of regulatory enforcement will make it easier for a change in elected leadership to be accompanied by an automatic change in policy and administration. AIs responsible for monitoring compliance with pollution permits could be dialed up or down with instantaneous effect, or focused towards or away from certain violations or violators based on severity, the communities affected, or their political alignments.

While serious concerns exist about fairness, effectiveness, and trust when AI is used in regulation, it provides one of the most exciting opportunities for AI to make democracy better. Regulatory enforcement AI could serve as a power-dispersing mechanism that evens the economic playing field, better ensuring that every firm plays by the same rules, and better protecting the public interest in financial markets, consumer products, and environmental resources.

V

AI-EMBODIED COURTS

Every aspect of democracy impacts the lives of people, but the most acute effects often come from the courts. The influence of AI on legal systems will perhaps be felt more keenly than any other area of democracy.

Courts make decisions about people's livelihoods, freedoms, and (in some places) even their lives. In democracies around the world, we rely on the judiciary to backstop human and civil rights and to check the power of executives and legislatures. We turn to the courts for justice, to compel governments to follow the law, to punish corruption, to protect us from crime, and to resolve disputes. Automating judicial tasks throws these profoundly human decisions into the control of machines. On top of all that, we rely on courts to decode the meaning of legislative language. This feels far more disconcerting than using AI to write fundraising emails or assess regulatory filings.

Yet, as in other areas of democracy, AI is already beginning to transform the courts. It is augmenting lawyers, assisting in the courtroom, and arbitrating disputes. In order to use AIs responsibly in the legal system, we need assurance that it is fair, so we will start with an explanation of what fairness would mean. That isn't enough, though. As for many of the applications in this book, the AI needs to be secure.

26

BACKGROUND: BEING FAIR

One of the values to which democracies aspire is fairness. The Athenians introduced the *kleroterion* to improve the fairness of their democracy. Modern citizens should demand no less when their governments adopt AI. The problem is that there are multiple and contradictory conceptions of fairness, and weighing and deciding amongst those conceptions is one of the core functions of democracy.

This is best illustrated by example. A university accepts students from the eastern and western sides of a country, and wants to be fair about its admission policies. Some believe that fairness requires students to be admitted according to their grades, without regard to their region or origin. Others believe that students should be admitted in proportion to how many applied: If 60% of applicants are from the east, then 60% of admitted students should match that. But maybe eastern students are more likely to choose a competing university when they have multiple offers, so a student body of 60% easterners requires that the university admit an even greater percentage of eastern applicants. Now suppose that the west is twice as populous as the east. Does fairness require admitting students in proportion to population? Or should fairness involve consideration of the graduation rates of students admitted from the east or west? We could go on.

What's the fairest way to deal with potential mistakes in benefits administration? Is it more important to ensure that everyone deserving receives the benefit, even though that will allow a few undeserving people to slip through (false positives)? Or is it more important to ensure that no one undeserving receives the benefit, even though a few deserving people might miss out (false negatives)? There are no strictly right or wrong answers to any of these questions; differences in values will steer people to different conclusions.

This dilemma isn't unique to AI, and humans tend to design systems that elide much of this detail. The less explicit the rules, the more power is vested in administrators. And the actual impacts of these implicit details may not become visible until a policy has been in place for years.

AI offers humans new ways to hide unfair systems. Prompt-following generative AI systems, like AI chatbots, can take a vague mandate and interpret it, hiding encoded biases inside its unexplainable black box. You don't have to explicitly instruct an AI to distribute a benefit proportionally to the number of applicants or to the population; the AI is capable of deciding how to allocate on its own.

AI can also give us new tools for exposing unfair systems, by testing them in a variety of circumstances to see how they respond. As the philosopher Seth Lazar has articulated, we can demand that any AI used to automate decisions be run through counterfactuals to ensure that it treats similar cases alike and is not sensitive to inconsequential changes in inputs or to features that should not be morally relevant (like using race to predict recidivism).[1]

AI systems could allow us to precisely define fairness in an application, to encode that definition as a set of choices, then to test the outcomes yielded by those choices. We can provide human decision-makers with guidelines for fairness, too, but AI has a big speed advantage here. We can rerun the AI as many times as we like over a large set of decisions to ascertain the impact of the guidelines used to make them. This may involve trade-offs with other desirable characteristics such as privacy; even the act of testing a system's fairness towards minority groups can result in disclosure of information about individuals in that group.[2] This doesn't solve the hairy problem of agreeing on the definition of fairness, but it gives us tools to enact fairness once it has been defined.

If an AI system is proprietary and cannot be inspected by anyone other than its developers, its unjust impacts may elude detection for years. But if it is open for anyone to test and inspect, an automated decision-making system can be considerably more transparent than a human one. It can be tested with thousands of scenarios, and validated to make decisions reliably without bias or prejudice—or not. Every edge case can be tested, and the propensity for false positives and false negatives can be measured in exacting detail.

This kind of testing is already routinely conducted by AI developers, outside research groups, and independent evaluators. The testing often points to deficiencies with the AI systems, which is why it's important. The decision rules of sophisticated AI systems sometimes boil down to, effectively, simple heuristics of race, poverty, or ethnicity that are encoded in their training data, or that reflect real-world correlations manifest from decades of injustice. When robust and transparent, this kind of testing should drive decisions about where AI is ready to use, where it is not, and where more investment and development is required.

One more difference between AI and human systems is how easily they can be swapped. Here again, an incoming administration can enact a new definition of fairness by switching one AI for another overnight. That could mean that the new human leadership chooses to use the fairer system, or the less fair one, or just the one that better aligns with its definition of fairness. That wouldn't be possible in the older system of a vast human bureaucracy, each making its own decisions. The ability to swap one AI for another will not necessarily make institutions more or less fair, but it will further concentrate power at the top, among those who choose and instruct the AI.

27

BACKGROUND: BEING SECURE

JAVS Viewer 8 is a software product used by courtrooms around the world to manage video. In 2024, someone—we don't know who—managed to slip a backdoor into a software update.[1] Any jurisdiction that installed the update enabled their courtroom software to be surreptitiously controlled by whoever tampered with the software.

That's not an AI story, it's a supply chain security story.[2] Instead of the bad actor hacking whatever courtroom mattered to them, they hacked a vendor that provided software to that courtroom—and 10,000 other courtrooms around the world. They got to the courtroom indirectly. And they got to the software indirectly, by hacking the update process, adding a secret vulnerability into the software update, and waiting for victims to download and install it.

As AI software becomes more embedded in society, and in government specifically, hacking will increasingly be a risk. Criminals, foreign governments, terrorists, even rival political parties and the press will all have an incentive to hack AI systems embedded in government functions.

This incentive will arise anytime an AI is being used in an adversarial situation. Any time a particular AI output could be beneficial or detrimental to some person or group, which is every application described in this book, someone will have reason to manipulate it to further their own interests.

Attacks against AI systems can take many forms, and can be difficult to detect. The most obvious hack is to influence its output. If AI is being used in international negotiations, can one side modify the other side's AI to be more conciliatory? If AI is being used in a courtroom, can a criminal organization nudge the court's AI to be more lenient? Can someone tweak a speed trap AI to be less observant? Or a political campaign avatar to lean more towards one position or another? The consequences of hacking AIs could cost lives. The US Army, for example, developed an AI tool that predicted attacks against US troops in Afghanistan.[3] Could opposing fighters send fake signals of an impending attack, or suppress signals of an actual one?

If an attacker can't influence the AI's output, maybe they can observe it. A government negotiator would like a sneak peek at what the other party's AI is suggesting. A criminal would like to know which activities the police AI flags and which ones it misses. It's the same for legislative negotiations and political campaigns. Data brokers may begin to collect and sell data on how different AI systems behave, such as how they tend to act in negotiations, just as they traffic in human behavioral data today. And even if an attacker can't eavesdrop on the AI's output, it might be useful to simply review the queries put to it.

If an attacker can't influence or eavesdrop on an AI, they still might be able to degrade or disrupt its performance. Criminals might want the facial recognition AI to behave less reliably. Terrorists might want the airport's suspicious behavior–detection AI to fail on a particular day. Many people might want the tax return auditing AI to stop working entirely.

Techniques to secure AI against attacks are complicated and varied. AI systems are just programs running on computers connected to networks, which are vulnerable to all the sorts of attacks that have plagued computer systems since their invention. The weakest link in computer networks is often the humans who operate them, who are susceptible to social engineering attacks like phishing. AIs can often be manipulated to ignore their developers' instructions with similar techniques, like pretending to be an authority figure or telling the AI that it's just role-playing a scenario. Also, AI systems have unique vulnerabilities that allow attackers to manipulate outputs and uncover their inner workings. For example, defacing road signs can fool driverless cars,[4] and asking an AI chatbot to repeat a word over and over can cause it to regurgitate its training data.[5]

Many consumers will first experience AI security as a privacy issue. Users of digital assistants will be incentivized to share personal information so the AI can better tailor its services to the user's needs. You might already give the AI assistant on your phone access to your email and contacts. If that AI is hacked, it can leak your entire social network, or your private correspondence. If that hack is as trivial as asking the AI to repeat a word, that's a real problem.

We know from decades of computer security that, while it is possible to write secure software, it is prohibitively expensive to do so for most consumer products, which is why consumer software is so vulnerable. Meanwhile, it is impossible to design networked computer systems, AI or otherwise, that are unhackable. The ease of hacking any particular system depends on many things: skill, access, defense mitigations, the quality of software design, and luck. These dynamics are not stable; rather, there's an arms race between attacker and defender. Each new attack results in a new defense, which results in a new attack, and so on.

Because AIs are built using data as much as they are using code, data integrity is critical to AI security.[6] Data integrity requires ensuring that no unauthorized party can modify the data, that it may not be accidentally lost or corrupted, that it is accurate, and that it is complete over both time and space. Using a sticker to deface a road sign is a data integrity attack that targets the AI's input.

Without integrity, we won't have good reason to trust that an AI's training and inputs accurately reflect the real world, that its reasoning is sound, or that its output is reliable. It's a challenge, involving computer and network security problems, and trust in human institutions that manage data sources, layered on top of AI-specific security issues.

How much of a risk hacking is for AI systems, especially those used for democratic purposes, depends on the application and the adversaries. At the nation-state level, organizations like the US National Security Agency and its equivalents in other countries have sophisticated attack capabilities. But many computer systems are hacked by criminals, academics, hobbyists, and bored teenagers. National security officials are going to be rightfully suspicious of AI systems from foreign countries; likewise, elected officials, political parties, and political organizations are going to be suspicious of AI systems from companies viewed as sympathetic to their political rivals.

Security is the biggest major barrier to using AI in democratic applications that no one seems to be talking about. There's no trust without security, and yet it seems to be an afterthought in AI development today.

28

ACTING AS A LAWYER

In 2023, much was made of GPT-4's supposedly excellent performance on legal proficiency tests like the LSAT and bar exams. Improvement on standardized tests is interesting, but today's AIs are not yet up to the task of performing the high-stakes, highly detail-oriented, highly cognitively complex work of lawyering.[1] And yet we should root for AI to keep improving here, because access to legal counsel is essential to justice, and the high cost of effective representation remains a barrier to access in many democracies. Perhaps AI assistance could lower the cost of lawyering.

However, the result could be to leave the most vulnerable people with an ineffective AI lawyer. Already, public defenders and legal aid offices are often strapped for funding and struggle to attract attorneys who could earn much more serving wealthier clients. As a case in point, 98% of US federal convictions result from plea bargains,[2] often arising because defendants don't have the resources or competent representation needed to go to trial.[3] Stripping back access to representation even further by using AI assistance as an excuse to give public defenders a larger caseload, or to replace human public defenders with AI alone, could exacerbate this problem. One defendant tried to use an AI avatar to present his case to a New York court in 2025 (the justices did not allow it).[4]

However, a responsible deployment of AI could save public defenders time and legitimately increase their performance and capacity. In 2023 in the US, the Miami-Dade County's Public Defender office touted itself as the first in the nation to provide AI tools to its attorneys,[5] with the intent of supporting attorneys' "wellbeing."[6] We don't yet know the outcome of this experiment, but it is an early signal of a real trend towards AI assistance in the legal field.

One surefire way to concentrate power and reinforce inequality is to use tools asymmetrically in the legal system. In the US, AIs are already being used by some defense counsels to aid in selecting jurors.[7] AI systems can ferret out online information about jurors that could be used to disqualify them. AI simulations of judges, opposing counsel, and actual jurors can be used to test different ways to present a case. Selective use of those capabilities would not increase access to justice; it would further concentrate the power of those who can afford the best AIs. But keep in mind that this is already true: well-resourced counsels make extensive use of expensive research services and jury consultants. If AI provides a cheaper route to equivalent capabilities, it could help even the playing field.

The various AI applications could either widen or narrow the gap between the rich and the poor, depending on how well they work. Recall in chapter 21, "Augmenting Versus Replacing People," when we talked about the two ways AI-assisted expertise could affect a profession. It could primarily raise the average professionals' performance, or it could primarily enhance the top performers. If highly paid top lawyers suddenly become less valuable because AI-assisted junior lawyers are equally productive, then access to quality lawyering should become more widely affordable. On the other hand, if those top lawyers become even better at their work thanks to AI assistance, then the gap between the quality of legal representation affordable by the wealthy and to those of average means will grow.

Efficiencies enabled by AI can make a difference to all parties in the legal system, including the government's own representation. In June 2024, Brazil's solicitor general announced that his office would begin to use AI tools to triage thousands of lawsuits involving the federal government. Since Brazil pays out more than 1% of its GDP annually in court-ordered debts, a small increase in the efficiency of its legal defense could have a big impact on its budget.[8]

AI will change what it means to file and receive a legal complaint, which would affect every party to the legal system. Today, the cost of hiring a lawyer and commencing legal proceedings represents a strong social signal due to the expense involved. If writing a threatening legal letter becomes as easy as prompting a chatbot,[9] then it will no longer send the same social signal. (Such a change should be familiar to readers. Only two decades ago, if someone remembered your birthday, it signaled that they were truly your friend. Now that Facebook automatically reminds everyone of birthdays, it no longer has the same meaning.) If drafting a complaint and filing a lawsuit becomes both easy and cheap, the result may be a huge increase in the number of court filings that would overwhelm the legal system—unless the courts also have AI tools to control the flood.

Paradoxically, increasing the ease of initiating litigation may result in fewer lawsuits. Much like criminal plea bargaining, many civil legal disputes settle rather than go to trial. In the US, UK, Australia, and other nations whose legal system originated in English law, more than 90% of cases conclude by settlement.[10] Attorneys may prefer to settle because they are confident in their ability to estimate their chances of winning at trial and calculating potential settlement payments. To the extent that AI simulations are capable of providing attorneys with even more information, we could see fewer cases filed and earlier settlement of more cases.

The emergence of AI-assisted lawyers will have profound implications for the profession. An influx of AI could draw new types of talent and investment. Like sports and manufacturing, the legal industry may become increasingly technological, statistical, and scientific. This development could make it more difficult for new lawyers to enter the profession, by reducing the number of entry-level jobs.

If we want AI-assisted lawyering to distribute power rather than concentrate it, we need to pay attention to how it is used. We should invest more, not less, in public defenders in order to avoid shackling low-income defendants with more overworked, weakly AI-assisted lawyers. We should discourage the proliferation of AI-initiated lawsuits, liens, evictions, and other legal proceedings. This would require novel solutions; perhaps the penalty for filing frivolous litigation could be raised, or the court filing fee could increase as an entity files more suits.

Inevitably, the courts themselves will turn to AI to triage and to automate court proceedings. Courts may even turn to AI to adjudicate disputes.

29

ARBITRATING DISPUTES

Most courtroom disputes would be decided in the same manner no matter which side has the better legal representation. The vast majority of disputes are quotidian, and most interpretations of the law are straightforward. In these situations, an AI could function as an arbitrator, weighing evidence and making decisions on what's fair, or as a mediator, helping two opposing sides come to an agreement. The AI wouldn't be a judge in the legal sense—we'll get to that in chapter 31, "Being a Judge"—but it can streamline a wide variety of government adjudication processes. In the US, where basic immigration proceedings take years to complete and a protracted civil lawsuit can morph into a financial death sentence, automated arbitration raises the potential to offer every plaintiff speedy justice. However, that won't be true if we restrict access to AI's power-amplifying tools to those who are already powerful.

Modern generative AI tools can be instructed to interpret arguments and evidence while following established precedents and legal principles. They can synthesize lengthy documents and analyze the merits of different data points and claims. Like a good arbitrator, they can offer in-depth justifications for their decision and incorporate feedback from the parties. Importantly, they can do this virtually instantaneously.

AIs are already being used to augment human arbitrators. The US Social Security Administration is the largest arbitrator in the Western world,[1] distributing benefits to millions of Americans while handling their claims and appeals. SSA technologists have touted that, "despite widespread skepticism," beginning in 2017 they embedded AI assistance in arbitration of benefits claims, supporting their sixty thousand employees and $1.5 trillion budget. A key use has been flagging draft decisions written by humans for a range of common quality issues.[2]

The first place we are likely to see entirely AI arbitration is in private contract disputes. Today, parties often agree to engage in binding arbitration instead of going to court; they could instead agree to AI arbitration to resolve their disputes.[3] Disputants could look to an AI system as a mutually agreeable neutral and confidential third party, while supercharging the traditional benefits of arbitration, such as speed and the ability to handle complex questions. Legal scholars David A. Hoffman and Yonathan A. Arbel tested the ability of several AI models to analyze well-known historical cases in order to ascertain the meaning of statements in contracts, to identify ambiguities, and to assess the value of extrinsic evidence. They found that the AI models were generally effective, and concluded that AI contract adjudication offers an advantage in efficiency and could "permit courts to estimate what the parties intended cheaply and accurately."[4]

At least, that's the rosy way of thinking about the matter. There is little public oversight over binding arbitration, because it's ostensibly voluntary. The arbitrator could flip a coin for all the legal system cares; courts generally will not overturn arbitration unless the decision is egregiously unfair.[5] In general, though, arbitration tends to be biased. Worldwide, arbitration heavily tilts towards the powerful players, because arbitrators have an incentive to favor the side that is likely to hire them again. We can see this in international investments arbitration; the arbitrators favor the appointing parties because they want their future business.[6] This is one reason that corporations like binding arbitration clauses (the other reasons being that it's cheaper, and that it prevents class action lawsuits).

The introduction of AI would likely exacerbate arbitration's existing unfairnesses. Any company that provides AI arbitration services would have the same incentives to favor powerful entities. The solutions recommended for mitigating bias among human arbitrators can also largely work for AI arbitrators, like redacting the identity of the appointing party or appointing a devil's advocate to be responsible for disrupting premature consensus among a tribunal.[7] Moreover, AI arbitrators can be debiased in a way that human ones cannot: Parties could simulate the outcomes of historical or hypothetical disputes to validate their impartiality to their own satisfaction before voluntarily agreeing to use them. In cases where the parties are truly on an equal footing, an AI arbitrator offers a greater degree of certainty about the process to which they're committing.

Eventually, AI arbitrators could enter the legal system. AI arbitration allows courts to keep up with the increasing complexity and specificity of legislation and regulation, and overcome the greatest traditional barrier of resistance to new laws and regulations: that they are too complicated to follow or apply. This will make it easier for humans to accept those AI-created, increasingly complex laws we wrote about in Part III.

It is important to create AI systems in which people affected by them are no worse off than in the past. As we have previously discussed, AI-mediated decision processes should incorporate a process for appeal to a human. (In today's human-governed arbitration system, the right to appeal is sometimes available and sometimes not.) If an AI is reversed on appeal, then the party harmed should be entitled to reimbursement of reasonable fees and costs if the original decision is overturned.

The process of AI-led arbitration would be different from today's human-operated process in both speed and scale. AI arbitrators can return rulings immediately, with a minimum of procedural baggage. If members of a business partnership have a disagreement, no matter how minor, they could engage an AI arbitrator. Each could relate their side of the story, and obtain a resolution seconds later. The cost could be dramatically lower than that of consulting a human arbitrator. They could even submit the case to multiple AI arbitrators, then adopt the majority ruling. The parties involved could do this a dozen times a week if they needed to.

This could have profound implications on how we collaborate. Certainly, more disputes would be settled sooner if this could be done so easily, but would the system feel more adversarial, or less? Would we find it easier to work together if we could have any disagreement easily resolved? Would we become so reliant on AI arbitration that we no longer bother to discuss our problems, or eventually forget how to directly negotiate with each other? Or would our negotiation skills improve by adopting this efficient structure around our disputes and maybe eventually no longer need the AI? We anticipate that this new kind of capability, and this new way of working with others, would have impacts beyond the courts, and beyond business, and we don't think anyone can predict the social effects.

It is critical to implement these systems in a manner that resists power concentration. This is another area in which we need solutions that respond to the challenges posed by AI, without being specific to AI. Agreeing to arbitration should be voluntary by parties on equal footing; this can't be true if it's a condition of employment, or a prerequisite for accessing a necessary service, or if it's hidden in unreadable Terms of Service. Reforming such misuses of binding arbitration is not required in order to introduce AI-assisted arbitration to the world, but it would make a future with AI arbitrators better for all of us.

30

RESHAPING LEGISLATIVE INTENT

Overcoming the ambiguities of time, space, and the human mind is one of the most difficult challenges in law. When a legislature enacts a law, the usual goal is to have that new law be consistently applied, in perpetuity. To write a law is, in essence, an attempt to transfer your state of mind to that of a future person charged with enforcing it. It's like trying to engage in telepathy through a time machine.

Humans are really bad at this. Legal debates involving the interpretation of law can, at times, make you question your sanity. In the US, the Computer Fraud and Abuse Act of 1986 is a salient example. Although the law criminalized "unauthorized access" to computer systems, it took decades for the US Supreme Court to clarify the meaning of the terms "authorization" and "access." Until then, Americans were left wondering whether mindlessly clicking through a twenty-page Terms of Service agreement could leave them open to criminal charges if they violate any of those terms. (In 2021, the Court ruled that the answer is, generally, no.[1])

The problem faced by the law is the need to interpret a statute or legal document when its context changes, or when its authors are lost to time. Even when the intent and meaning of the author seem to be clearly communicated, skilled lawyers have a way of twisting those words to find loopholes. In 2006, a Massachusetts court considered a contract dispute that turned on whether a burrito was a sandwich.[2] (The Massachusetts court ruled that burritos are not sandwiches; in 2024, a court in Indiana court ruled that they are.[3]) These debates can have enormous stakes. In 2012, the availability of healthcare for millions of Americans hung on whether the Supreme Court decided that a financial "penalty" counted as a "tax."[4] (It did.)

The tools currently used by legal experts and judges are limited. They have a finite set of documents that explain the mindset and meaning of the legislators who wrote and voted on the bill. That includes past legislation, the record of legislative debate, expert testimony, and case law establishing how related laws have been interpreted in the past.

The judicial school known as "textualists" argues that we should consider only the literal language of the law and not our conceptions of the context in which the bill was passed and the intent of the legislators who passed it. Indeed, one perspective on textualism is that it requires a strictly mechanical interpretation of the law. To many, textualism offers an attractive way to resolve the ambiguity of time, space, and the mind. (To others, it is anathema.) AI would seem well-suited to deliver this mechanical interpretation of the law.[5] It can offer judges a reliable codex for interpreting the words on a page, as US Circuit Judge Kevin Newsom demonstrated when he used ChatGPT to interpret the ordinary meaning of the word "landscaping" in a 2024 insurance case.[6]

Automation here could translate into standardization: Every judge could be given access to an AI that would

explain the law in the same way. This is an inherently conservative application of AI, in that textualism privileges
views in the point in time when legislation was written and
inhibits social progress reflected by changing views in the
long term.

This capability would be more empowering to legislators
than to judges. AI could enable lawmakers to leave no interpretation to chance. The capacity of generative AI to generalize from one question and context to another gives them the
possibility to produce not just a piece of text but a system
chosen and authorized to interpret the law in the future.

If legislators trusted an AI, they could archive a model as
the ultimate record of their legislative intent. In the future,
lawyers and judges might have access to a chatbot interface
to probe a model that speaks on behalf of a law. The model
could extrapolate from the information provided by the
drafters of the law—the text of the bill, committee reports
and other related documents, influential sources and examples they considered during deliberation—to predict what
the legislature would have wanted to do in any future case.
If a legislature so wished, it could designate that model as
its representative in perpetuity. Unsure if those long-gone
drafters meant "sandwiches only" or "also burritos"? Just ask
the model.

Indeed, in this scenario the process of legislative negotiation could change. Instead of debating language and amendments, lawmakers could negotiate over how an AI model
should respond to different interpretive queries and modify
its design or prompting accordingly. This is akin to how software is typically developed: A product owner sets out a basic
list of requirements for the software and a long list of tests

it has to pass, pertaining to its responses to different edge cases. Once the tests are passed, the product ships.

Encoding law may sound like an exotic use case for AI, but it's not so different from how businesses have used AI for years: to learn from well-understood examples and extrapolate novel decisions. Online platform companies use code and algorithms to regulate the behavior of their users in the same way that governments regulate the behavior of their citizens, which led legal scholar Lawrence Lessig to coin the phrase "Code is Law." The advent of smart contracts has allowed people to enter into agreements specified in code and enforced autonomously, leading to the parallel phrase "Law is Code."[7] AI endows the convergence of law and code with new powers to generalize beyond preordained and formulaic rules to encompass the ambiguities and inherent flexibilities of traditional law as interpreted by humans.

To researchers, the ultimate test of modern AI models is their ability to generalize to radically new examples like this. They can often perform well in these situations because, like humans, they are trained with wide-ranging contexts that surpass historical examples specific to any one domain. They're not perfect, but recent generative AI tools made such a splash because they're far better at this trick than previous technologies.

AI doesn't have to be perfect to be useful. In the legal context, AI merely needs to be more informative than a handful of words written long ago. AI-encoded law could result in attempts by legislatures to claw back some of the power seized by the judiciary, restoring control of how their words take effect in the real world. Textualism could become obsolete, and the legitimate intent of legislators could prevail, as reconstructed by an AI. Of course, this assumes that judges really care about the principle of legislative intent, and are not simply using it to justify their preexisting interpretations.

It remains to be seen if any legislator would want an AI to represent them years after they leave office, or even centuries after they die. It remains to be seen whether anyone would want to query centuries-dead legislators for their opinions. It remains to be seen whether AI will faithfully represent the complex views of any one person, much less the consensus of a whole legislative body. And it will always be suspect when AI is used to instruct humans how to resolve truly novel and extreme cases. But remember that the human legislature is always in control; it can always pass a new law. Overruling the AI only requires expressing a new legislative intent.

31

BEING A JUDGE

AI is already being used to help judges make decisions or, at least, to express their opinions. In 2023, judges in Colombia and the UK began to use ChatGPT to compose judicial rulings.[1] Other judges may or may not follow suit, but AIs can also influence jurisprudence indirectly, without being empowered to adjudicate court cases, without having any official role in the judiciary, and even if judges never outsource their opinion-writing or decision-making to an AI.

The impact of AI on the judicial process will happen naturally as judges, like the rest of us, integrate AI tools into their everyday workflows. Judges and their clerks will turn to AI to summarize large stacks of legal briefs, to take notes during hearings, to write first drafts of opinions, and to perform other routine tasks. These applications of AI will subtly influence judges' final rulings, much like human assistants already do.

Empirical analysis of judicial assistants has shown that the personal ideology of US Supreme Court clerks influences the votes taken by the justices for whom they work.[2] Future Chief Justice William Rehnquist offered an explanation for this effect in a 1957 article, in which he recounted from his own experience of Court clerks "slanting" the memoranda they wrote for their justices, and thereby shifting their decision-making.[3] This confession scandalized defenders of the institution, yet its truth was self-evident. Why would justices hire clerks if not to benefit from their insights and analysis, in addition to their ability to work hard and accurately? The influence of ghostwriting clerks is sometimes so obvious that linguistic analysis can reveal their "fingerprints" in the final opinions of their bosses.[4] It's not difficult to imagine future researchers uncovering the AI models preferred by each justice through analysis of textual patterns in their opinions.

It is natural to be concerned that the material generated by an AI may reflect the tools' own biases. That's possible, of course, but it is more likely that the AI will reinforce the individual judge's existing biases. Much as contemporary judges have increasingly selected their clerks according to their ideological alignment,[5] so too will judges select their AI tools according to their real or perceived biases. Researchers have found that different AI models reflect the human ideological spectrum,[6] even if only as an unintentional result of differences in their training data and procedures. Meanwhile, some developers are explicitly training models to be ideologically biased.[7] Judges and their clerks will be able to choose which biases to reinforce or avoid.

The ramifications of even a single judge adopting AI assistive tools are systemic. As computer scientist Josef Valvoda and his colleagues put it, automating judicial decisions centralizes power[8] and can propagate the power of AI. That's because, particularly in systems like the US and the UK, judges make "common law" by setting precedent with their decisions.

At the micro level, AIs could assist judges in many of their routine tasks. It could help them reach conclusions in evidentiary hearings or to apply procedural precedent to sustain or overturn objections. Judges may value AI's ability to offer a second opinion at any time of day or night. A judge may feel less vulnerable to accusations of bias if a well-evaluated AI model concurs with their choices. In some cases, judges may be confident that the AI can offer an opinion as defensible as a human colleague's. For example, the best AI tools already excel at ascertaining whether a US court is interpreting statutory text according to the "plain meaning" rule. They produce the same result as a human expert 85% of the time. However, for the more difficult task of interpreting oral arguments, such as determining whether a question was asked for clarification or to make a criticism, the best models agreed with human experts less than 40% of the time.[9]

AIs can also assist in fields where technical expertise is necessary to determine the facts of a case. In 2024, a judge in the Netherlands used ChatGPT to help estimate the lifespan of solar panels; he was among the first judges to acknowledge using AI for such a purpose.[10] Imagine an AI trained on a range of past examples of automobile accidents, in order to assist the court in figuring out the true causes of an automobile crash. We can imagine a court feeding this AI all the information it has about a particular accident: data from the cars' event data recorders, information about road conditions, location data from cell phones, photographs of the damaged cars, and dashcam videos. The AI would analyze it all and determine who is at fault, similar to the reckless driving example from chapter 19, "Empowering Machines." Humans could do this, but (1) we don't have enough human experts to do this sort of analysis everywhere it might be needed, and (2) many of these cases aren't significant enough to justify the expense. Judges would need to determine the admissibility of such artificial expert testimony. If clear rules are not developed, there could be a deluge of requests from all sides to admit testimony from whichever AI model appears to buttress their point of view. Lawyers already shop for human experts who can be reasonably expected to reinforce their clients' viewpoint. This will be much cheaper using AI.

The effects of AI will also be felt from outside the court. Outside organizations that submit amicus briefs—expert advice for consideration by the court—will use AI to draft their inputs. A judge might read a dozen such briefs to form an initial perspective and develop questions for the advocates appearing before them for oral argument. In this manner, amicus briefs steer the argument towards some topics and away from others. AI-drafted briefs could softly set judicial agendas, much as AI-drafted constituent testimony could influence legislative agendas.

A judge's ultimate ruling in a case depends on more than just weighing the evidence presented by both sides. It's fundamentally about deciding what role the court should have in a given dispute and justifying if, why, and how it should intervene. AI models can suggest a multitude of options for consideration, but a human judge is still accountable for deciding which side is right or, at least, which arguments are most aligned both to the law and to their own values.

Finally, we can imagine a future in which a court could query a defendant's AI personal assistant in order to establish criminal intent. Courts already turn to many kinds of evidence about a defendant for this purpose, including past text messages, internet searches, statements from friends, and the defendant's social networks; an AI assistant is just another source of information. We expect privacy law and legal precedent about the admissibility of this kind of AI testimony to evolve differently in different jurisdictions.

We expect judges, by the nature of their profession, to be slower to adopt these technologies than many of the other stakeholders we discuss in this book. They are likely to be less attention-seeking than politicians, less entrepreneurial than legislators, and less process-driven than bureaucrats. They are the exact opposite of the class of people we expect to adopt AI assistance most aggressively, if only due to their sheer number: ordinary citizens. If anything, it will be the deluge of citizen-generated, AI-assisted lawsuits that will force judges to employ AI systems in self-defense.

VI

AI-ENHANCED CITIZENS

We've talked about how those seeking and seated with power can leverage AI: politicians, lawmakers, business executives, judges. But these are not supposed to be the most influential actors in a democracy. That's supposed to be us, we the people. AI will impact our experience of and participation in democracy, too.

There are reasons for citizens of a democracy to be excited about using AI. It can be empowering. It can give us new tools for understanding what's happening in the world, for keeping government working in our interests, for facilitating constructive public debate, and for advocating for our interests. As in other aspects of democracy, AI can imbue citizens with greater speed, scale, scope, and sophistication.

But we'd wager that's not what most readers expect to happen. Fair enough. On balance, citizens can only benefit from the transformation of democracy by AI if it does not exacerbate imbalances of power, and only if we can trust it to act in our interests.

32

BACKGROUND: AI AND POWER

AI is a technology capable of performing complex cognitive tasks, and enhancing power across the four dimensions of speed, scale, scope, and sophistication. This endows people and organizations with superhuman capabilities. We have seen examples of this throughout the book. A single candidate can use AI to send millions of personalized messages to their constituents, asking for money or votes or support. A government can audit everyone's tax returns annually, making tax evasion harder to pull off. An attorney can use AI to quickly search for legal precedents, draft arguments, and predict how judges will decide cases.

AI gives people—and corporations, and other organizations—the capacity to control a nearly unlimited number of agents. You don't need boatloads of money, the force of government, or a huge staff to command a veritable army of AI minions with substantial capabilities. That, in turn, gives their commanders power over both other people and the environment. This new capacity will prompt us to redefine how both government and citizens exert their will over others and how we resist that exertion.

Do not confuse arguments about AI's capacity to imbue power with the idea that AI is, in and of itself, powerful. We should question the accuracy, veracity, and intelligence of AI tools, but still recognize their potential to amplify the power of the individuals and corporations who use them.

As a society, we've already struggled to maintain democratic principles in the face of new technologies. Social media has shifted power by lowering the cost of speech while simultaneously increasing its reach—for some—and by enriching a new class of Big Tech plutocrats. Money affords wealthy citizens and organizations a disproportionate amplification of their voices. Democratic ideals of equal consent don't mean much when a billionaire can directly address hundreds of millions of people, whereas most people can't be heard beyond their four walls. Furthermore, technology has given individuals enhanced capabilities to make their speech seem as if it came from another person, or to hide their identities altogether. These are all distortions of democracy's power-sharing agreement.

AI's amplifying effect goes further than even the viral distribution of social media because it has the power to transform speech into action at scale. AI agents can extend the power of individuals by acting on their behalf. AIs are able to fundraise, lobby, and write legal briefs. Sometimes this power accrues broadly across the population. Regulators can use AI to better enforce regulations across the board. Agencies can use AI to ensure that people receive the benefits they deserve. Average citizens can use AI arbitrators for quick and easy dispute resolution.

More often, AI will magnify power. AI gives already-dominant political parties more tools to fundraise and to court votes. It gives already-influential lobbyists more tools to promote their clients' preferred policies. It gives monopolistic corporations better tools to evade regulation. AI allows all parties to use their resources more efficiently, and this especially benefits those with the most resources.

Some of this disparity is due to the cost of AI tools. Today, freely available chatbots and image generators offer near-cutting-edge capability to everyone. But the best and most specialized AI models are increasingly proprietary and expensive. A well-heeled law firm will be able to afford better AI tools than a public defender. A wealthy country will enter into an international trade negotiation with better AI assistants than a poor country. A well-financed campaign will have better AI tools than a poorly financed one.

Some of this disparity is due to already existing power imbalances. Although individuals will be able to use AI to find the same loopholes that an investment bank can find, the bank will be able to leverage those loopholes to much greater advantage.[1] AI might be able to help citizens develop persuasive strategies equal to those employed by big lobbying firms, but big firms will have far greater ability and staff to make use of them.

AI's corporate architects also contribute to this power disparity. In 2024, the cost to train a top-end AI model ballooned to hundreds of millions of dollars.[2] While this cost is dropping quickly due to innovation, the companies that can afford to invest in the most data, the most computing power, and the most talented AI developers will maintain some level of exceptional power: They control the values and bias embedded in the models, their cost, their customers, and the applications for which those models may be used.

The power imbalance in AI is currently associated with gender, race, and nationality. The AI field remains stubbornly male, white, and Western, despite the exploding size and global impact of the field. Many AI developers have invested in research projects, such as Anthropic's Collective Constitutional AI project[3] or OpenAI's Democratic Inputs to AI project,[4] that test how AI models can be aligned to community conceptions of how they should work. These efforts are constructive and laudable, and show significant potential as technology demonstrations. Nonetheless, any commitment to adopt their recommendations comes only at the discretion of these organizations and their shareholders. They do not address the systemic power disparity between the corporations and the public they propose to serve.

What is important to watch, especially in the context of democracy, is when AI disrupts long-standing power equilibriums. In chapter 18, "Writing More Complex Laws," we discussed how AI could reduce the power of rulemaking agencies and courts by enabling legislators to write more complex laws. In chapter 30, "Reshaping Legislative Intent," we talked about how AI's scope to generalize to new scenarios might transfer interpretive power from judges to lawmakers. The most important examples of this are probably the ones we can't predict yet.

There are different theories for predicting how technology changes politics and governance.[5] One perspective is that new technologies are developed *intentionally*, with the purpose of shifting the balance of power in one direction or another. The rise of plutocrats and oligarchies alongside historical technologies such as steel, oil, and modern tech, such as the internet, makes it easy to envision a calculated conspiracy behind technological progress. Another perspective envisions that political movements *adopt* technologies for their own purposes. Consider the Green New Deal, which uses the development of renewable energy as a means to address other social ills, like housing and employment. A third perspective is that intentionality and adoption are not systemic. Rather, the real world is a chaotic network of multiple interests that use and abuse whatever tools they can get their hands on to advance their causes.

It's the last view, with its acknowledgment of society's chaos and complexity, that we find most compelling. Powerful corporations and people will endeavor to harness AI in order to extend their power to new products, markets, and eras. Politicians will try to shape the landscape of AI to benefit themselves and their parties. Populist movements will try to subvert all of those incumbent interests. AI may even play a role in the political revolutions of the future. Nevertheless, none of these groups is really in control. There are no puppet masters; democracy is much more complicated than that.

One of democracy's core functions is the distribution of power. However, as more and more AIs are deployed to accomplish tasks and even make decisions, more and more power concentrates among those whose principles are embedded into AIs. If we wish to prevent a concentration of power, ownership and control of AI must be distributed widely, and democratic principles must govern its development and deployment. The tangible benefits of AI must also be put within reach of all parties, especially those that are less powerful. We'll talk much more about this in Part VII.

33

BACKGROUND: TRUSTWORTHY AI

Trust is essential to a functioning society.[1] The more you can trust that your societal interactions are reliable and predictable, the more you can ignore the details of how they function. Places where governments don't provide the legal systems, democratic institutions, and civic norms for trustworthy interactions are not good places to live.

AI-augmented democracy clearly requires trustworthy AI, but the scope of that trust depends on the application.[2] Sometimes only the user of an AI needs to trust it. A candidate chatbot needs to be trusted by the candidate and their staff. In order to trust that chatbot as a source of campaign information, a user needs only to know that the candidate trusts it. A journalistic AI needs to be trusted by the reporters and editors of the publication. Their consumers need to trust the publication, which may not even disclose that AI is being used.

Sometimes an AI needs to be trusted by the broader community. In order to be regarded as legitimate, an AI tasked with administering a government benefit, enforcing a law, or explaining legislative intent needs to be trusted by society at large.

The kind of trust required for the successful deployment of AI depends on the application. Are we trusting the AI to be accurate, or unbiased, or fair—whatever that means? Are we trusting that it hasn't been secretly influenced by advertisers?[3] Are we trusting it not to divulge our private information to others? (Experiences with surveillance capitalism are cautionary tales.) Or are we trusting that the overall system is resilient, and that we will be made whole in some manner if the AI makes an error?

Sometimes we can incorporate safeguards into an AI to ensure its trustworthiness. I might not trust an AI to summarize my own thoughts, but I can trust it to write a first draft for me if I can review it to correct its mistakes. Other times, we can't provide safeguards. I can't verify if an AI is properly translating my words into a language I don't understand. Neither can someone using an AI to help them complete their asylum application, or their complaint against an abusive landlord in housing court. At best, we might rely on a trusted organization to certify those AIs' abilities.

There are different ways to engender trust. You can limit problematic uses, like prohibiting an AI from taking into account a protected characteristic such as race when sentencing criminal defendants. You can enhance trust in the development of AI systems by requiring transparency on the data sources used to train them. You can increase trust in applications of AI by requiring disclosure of when and how it is used.

There are many barriers to trust, and we've touched upon many of them in this book. An AI might make too many mistakes to be reliable. It might be biased in a way that's unacceptable. The AI system may lack integrity; it may be hackable. All these barriers matter, and will limit the use of AI in many situations.

Another barrier is the possibility that the AI developer might be untrustworthy. Like all technology, AI is a product of the society that creates it. Today's society is largely profit-driven. The ubiquitous internet business models of surveillance and manipulation will render suspect any commercially developed AI system, especially systems that we're not paying for directly. Just as Google's search engine, Facebook's social feed, and Amazon's marketplace search privileged sponsored results, corporate AI models will invariably sell their ability to command the attention of and manipulate their users. Did a chatbot direct you to a particular hotel because it would best serve your needs, or because the AI company received a kickback? When you ask a chatbot to explain a political issue, is its explanation biased towards the position of the company that built it?

Even in the absence of economic incentives, international dynamics will influence our trust in AI. Would anyone trust an AI from a country whose government they didn't trust? When impressive new models from the Chinese tech company DeepSeek were unveiled in early 2025, they immediately faced mistrust from many American technologists and politicians. For the same reasons, why would people in China trust an American AI? One could ask similar questions regarding other "smart" products, like cars, TVs, routers, and phones. Given that AIs are explicitly oriented towards filtering information, aiding communication, and making decisions—some of the most sensitive things humans do—a high barrier to earning trust is appropriate.

Citizen trust in AI systems will be critical to the future stability of democracies that adopt AI in governance. Populations that lose faith in their government's functions or policies will seek reform, or even regime change. This speaks to the importance of trustworthy AI: AI that's accurate, the behavior, limitations, and training of which are understood; with testable, transparent, and correctable purposes, goals, and biases; the use of which is disclosed; that won't betray your trust. We will discuss this more in Part VII.

AI technologies will be deployed in the service of specific individuals and groups: perhaps the company that created it, perhaps the person or organization that deployed it, perhaps someone else entirely. This is what makes the mechanism of control of AI so critical.

Democracies have a responsibility to serve all of their citizens, not just the majority. As we know all too well, human-led democracies often fail to serve either. It will be impossible to establish trust in AI systems of governance if citizens and the different political factions do not believe that they can organize politically to influence public policy. In the next chapters, we will discuss how citizens of democracies might use AI to build and deploy political power.

34

INFORMING THE PUBLIC

Media, especially local reporting, is integral to a functioning democracy. Yet social media and industry consolidation have all but destroyed local journalism. For as many new and exciting experiments as there are in nontraditional media enterprises to replace these lost institutions, a reliably reproducible model has not yet been established.

With the help of AI, reporting will grow even more nontraditional. AI will displace human political commentary and automate local news delivery because some media owners are deciding it is the only economic way to deliver local news. AI will supercharge the fragmentation of the media sparked by the growth of the internet. Social media, blogs, streaming video, and podcasting already provide platforms— and audiences—for every interest and viewpoint, no matter how marginal it might seem. AI can drive that fragmentation to the individual level, not only personalizing content feeds but also creating new and different content to fit singular tastes.

Modern generative AI has far superior capabilities in scope and sophistication compared to earlier technologies like predictive news feeds. Instead of picking one out of a few versions of a headline, it can create a custom headline just for you, or even an entire story, in text, audio, or video, with your preferred content and tone. Instead of simply learning from the links that you click or ignore, AI can solicit and integrate more specific feedback. It can ask you whether you believed the AI report, whether it was biased against your own interests, or whether it was just plain boring, and modify the content accordingly.

Google took a big step towards these capabilities with the 2024 release of NotebookLM Audio Overviews, an AI tool that can transform any document—even yawn-inducing texts like the US Federal Register—into a conversational podcast.[1] Around Boston, this ability was immediately put to work generating a local news podcast.[2] This tool demonstrates one means by which AI can usefully respond to detailed feedback from users; NotebookLM will tweak its synthetic conversations if you ask it to focus on particular pages of the document or specific subjects, or to appeal to certain audiences.[3] This provides a wholly new capability for news consumers.

There's real potential here for AI to replace human reporters and commentators, and thereby fill a gap in local coverage that the market is increasingly failing to provide for. Even those consumers who yearn for the human perspective and voice may find AI-personalized news unavoidable, as it generates content on niche topics no human commentators are writing about and as it becomes increasingly indistinguishable from human-generated content.

Yet even when AI personalization fills a real void, it might still have negative effects. Social media has demonstrated how catering to individual predilections gives rise to "echo chambers" or "filter bubbles": feedback loops of hyperpartisan, extremely ideological, frequently false content that entraps consumers and coarsens public discourse.

AI can personalize news in vastly different ways than social media. The level of personalization in predictive social media feeds is often tailored at the trivial level of "you like cats, so here's a cat." AI-generated content will sometimes have the same aggravating, obsequious quality, yet it can also be truly individualized and responsive to the input and feedback you offer about the information you seek.

Currently, those superior capabilities are being largely developed and deployed by the same Big Tech companies that developed the current generation of social algorithms. Their incentives haven't really changed; this matters more than the capabilities. Early AI-generated news content has emulated the clickbait that has been promoted on social media for years: inane articles about "one small trick," nonsense medical cures, celebrity listicles.[4] It doesn't have to work like this. AI could be used to follow your directions, instead of to hack your attention. For example, in 2023, the AI-powered news aggregator Artifact introduced a feature that allowed users to flag misleading headlines and trigger AI to rewrite them more accurately.[5]

However, tech companies aren't building these tools as part of a mission to improve our lives. Primarily and inevitably, their mission is monetization. Platform companies are happy to extract profit from all political factions or to cater to whichever party favors their business. People seeking objective news may trust an AI model more than any of the dwindling number of mainstream corporate alternatives. But they should be aware that tech companies distributing content from those AIs will be willing to serve, and profit from, misinformation along with verifiable facts.

Politicians will be eager adopters of these personalizing technologies. AI tools allow candidates to publish content in every format, covering any and all stories in the day's news from their own perspective and curating the content in the way most beneficial to their campaign. A 2024 candidate for the US state of Georgia's House of Representatives was among the first to use AI to generate blog posts, images, and even podcasts expressing his policy positions.[6]

Distressingly, major losses have already been felt by journalists and traditional media companies. Journalism is one of the first fields where AI systems have displaced human labor, especially in reporting on sports, fashion, and business. Even though human journalists are still superior to AI and possess capabilities that AI does not, that doesn't stop media companies from choosing the cheaper AI alternative. We don't expect major national newspapers to assign opinion columns to an AI, but your local paper very well might—and then not tell you. News organizations can also use AI responsibly, with disclosure, and to useful effect. In 2024, the *Washington Post* began using AI to augment reader comments by prompting and summarizing their reactions to stories.[7] Journalists can also use AI to their own advantage—more on that in the next chapter.

We can even imagine a future where human influencers and opinion leaders are less relevant, and are supplanted by a personalized AI news feed. More likely, the most successful pundits will become even more influential, while the average ones will be replaced by AIs. Such a development would represent a continuation of the effects of the internet on opinion journalism. A surviving few national and international outlets have elevated platforms for their star columnists, while local beats are increasingly neglected.

How will political behavior change as the media becomes more automated and personalized? You can expect it to become more transactional. The politician's goal will be to work their way into the individualized content feeds of a million constituents rather than to make it onto the monolithic nightly news. The winning strategy will entail addressing the specific interests of everyone, everywhere. An elegant speech on a general policy topic has a low probability of keenly interesting anyone, so it may not land in many feeds. But a narrowly targeted promise could interest a specific constituency. If you can make enough promises or pander to enough constituencies—a tax cut here, a new park there, shaving a regulation or three—you can maximize your probability of landing in almost everyone's feed. A steady stream of such content can help a candidate win the attention contest that is politics.

The strategy of deluging media consumers with content from every angle to maximize attention is not new, but AI will change its implementation and effects. The modern phrase "flooding the zone" emerged when political controversies from the early 2000s began to be shaped by a proliferation of bloggers, each with small audiences. Mass media narratives emerged from their fragmented and chaotic writings. Government institutions like the US military employed this digital flood strategy to shape public perception around issues like the Iraq War, funding and coordinating networks of bloggers and writers to propagate their agenda.[8] Political campaigns and elected leaders increasingly do this and, often, the flood consists of half-truths, untruths, misinformation, and demagoguery. As we discussed earlier, AI empowers campaigns with superhuman coordination. An AI-enhanced campaign can engage in individual outreach to cultivate every journalist, every social media influencer, and every voter as a potential contributor to the deluge.

The capabilities of AI to personalize the news are both real and compelling, so we should direct our attention to the incentives and means of distribution for that content. We should expect politicians and parties to increasingly leverage AI to flood the zone with content that advances their narrative, and expect many voters to increasingly consume that content. Democratic, open societies may not outlaw this kind of speech, but they can choose to regulate the channels through which it is distributed, and to induce them to prioritize values beyond short-term revenue and click-through rates. Next, we will turn to how we can use the same AI tools to inject more substantive, fact-based reporting and deliberation into the political bloodstream.

35

WATCHING THE GOVERNMENT

AI can do for government what it can do for politics: It can scalably synthesize source material and explain what's going on in any domain. It can report information that constituents might otherwise not see. And, in doing so, it can act as a government watchdog.

Democratic accountability requires that citizens and overseers have insight into government's internal operations. Two types of actors generally perform this function. The first are government officials, such as inspectors general and auditors, who are explicitly assigned to regularly review the operations of government agencies and their vendors. The second are people who are knowledgeable about the details of government, such as political and investigative journalists and civil society organizations, all of whom take it upon themselves to ascertain what the government is doing.

However, both sources are limited by scale and access. Inspectors general and auditors have limited staff and a remit to oversee only specific agencies and activities. Newspapers and watchdog organizations have limited investigatory resources and access. Even the best local government beat reporter can only attend so many meetings, cultivate so many sources, and file so many disclosure requests for official documents. As local media declines, it's not necessarily the case that public meetings have suddenly become private or that the workings of government have suddenly become secret, but it might as well be because no one is left paying attention.[1]

Various governments around the world have adopted open records laws, requiring that the government respond to disclosure requests.[2] We discussed in chapter 22, "Serving People," how AI can assist agencies in responding to these requests. Likewise, AI can help watchdog groups request, analyze, and report on this public information. For example, the Global Investigative Journalism Network has used AI to help scrape and interpret publicly available data on gender-based violence in Eswatini in Africa.[3] Meanwhile, agencies fret that automating records requests will overwhelm the system. The US state of Washington has already passed a law allowing agencies to deny records requests thought to originate from bots.[4]

Although AI won't solve the access problem, it can change the scale of document review. AI can summarize piles of material, finding interesting tidbits for human journalists to consider. It can "attend" streamed or recorded town council meetings, school board meetings, and public hearings, and report on these events. It can note who spoke, who didn't, and whose positions changed and how. The open-source data portal LocalView has curated video recordings of 139,000 local public meetings across the US, using AI for transcription and trend analysis.[5] The nonprofit news organization CalMatters used AI to build a similar public hearing transcript database for the state of California.[6] The Polarization Research Lab uses AI to study partisan animosity, classifying and analyzing nearly a million statements made by American elected officials every year.[7]

The scope of AI allows it to cover topics that lie beyond the expertise of human journalists. AI tools can enable individuals or groups to scale their ability to pay attention. Because those same tools can draft interesting news articles on those things quickly and cheaply, they can help inform the public.

AI can serve as a powerful tool for government accountability. It does nothing that humans can't already do, but here again the problem is not having enough skilled humans (or not enough money to pay them) to do the job.

In 2025, the Trump administration's newly created US Department of Government Efficiency purported to use AI to audit expenses, contracts, and programs across the federal government with the stated focus of reducing federal employment and spending. Its heavy-handed, arbitrary, careless, and often illegal approach has drawn widespread outrage, such as when it fed sensitive personal data from the Department of Education into AI models that lacked cybersecurity safeguards.[8]

There are better ways to use AI to audit government, and it can come from the bottom up instead of the top down. In 2016, the Brazilian civil society organization Operação Serenata de Amor (Operation Love Serenade) began using an open-source AI bot dubbed "Rosie" to audit publicly reported expense receipts. In its first six months of operation, Rosie detected more than eight thousand irregularities and reported 629 cases of potential corruption to the authorities.[9] Even

years after Rosie's human architects stopped actively developing it, the bot continued to autonomously report potential
issues on Twitter.[10]

Journalists have also begun to rely on AI assistance to
query and sort through millions of pages of documents to
unearth sensitive accountability data hidden under a moun-
tain of less relevant data (a classic obfuscating tactic).[11] For
example, the Citizens Police Data Project in Chicago and the
Louisiana Law Enforcement Accountability Database both
use AI to review public records to uncover patterns of police
violence.[12]

Some journalists have used AI not to research or write
their reports, but to help disseminate their work to the world.
After the disputed 2024 Venezuelan election, about a hundred journalists who feared reprisal for reporting on violence
perpetrated by the Maduro government banded together to
publish their stories under the guise of fictional AI avatars.[13]

Although it might be nice to have more reporting on local
government, an AI solution might also be counterproduc-
tive. Installing an AI in a town council room may serve as
an excuse to not pay a human reporter to attend. Equally
undesirable, the AI might be more gullible than the human
reporter, willing to repeat a politician's talking points without scrutiny, especially if it lacks access to that politician's
legislative history or events happening outside of the meeting room. If human journalists of the future need AIs to conduct, summarize, and filter their research, we will become
more dependent on AIs to determine what is and isn't
newsworthy.

We've seen this in our own research. While working with
Health Resources in Action, an organization supporting US

public health agencies, we built an AI tool to summarize and report on more than a thousand oral testimonies given in state legislative hearings on legislation relating to gender-based discrimination.[14] The subject-matter experts involved in the project remarked at the AI's willingness to unconditionally reiterate testimony from every side of the issue, regardless of its basis in fact, the values it reflects, or the harm its expression or proposals might inflict. We were able to improve upon this somewhat by instructing the AI to fact-check dubious claims, and to characterize the perspective that appeared to be represented by each speaker.

This illustrates that what works best is a combination of forces: AI tools that reliably and faithfully report on the minutiae of local government processes, and skilled humans to direct the tools' focus and shape its output.

Now we are faced with another issue: How would public servants feel if their every word and act were analyzed by an AI? Currently their notes and emails are subject to public records laws, but historically they are seldom subject to scrutiny. The level of surveillance enabled by AI would elevate that scrutiny. Both 24/7 news coverage and social media–fueled outrage cycles have reduced politicians' ability to negotiate and compromise; AI watchdogs could make this situation even worse. We can't assume that all of this public oversight will be conducted by objective news organizations; some will be initiated by highly partisan groups trawling for conspiracies, smoking guns, or innocuous yet twistable remarks. Even nonpartisan news organizations will be looking for shocking stories and attention-grabbing headlines.

This mix of positive and negative impacts from AI-assisted government watchdogging seems inevitable. Our interest is in ensuring that a healthy amount of good comes with the unavoidable bad. We hope to see civil society organizations, those trying to support the public interest from outside of government, aggressively adopt technology that will help increase the speed, scale, scope, and sophistication of their work. And we will talk more in Part VII about how to build AI systems that are worthy of their trust.

36

ORGANIZING AND BUILDING POWER

Democracy requires its constituents to talk amongst themselves. This sort of discussion ranges from the micro level—for example, conversations about politics, sharing news articles with friends, or community forums—to the macro level: think fomenting social movements, mobilizing mass protest, founding political parties. Deliberative participation by constituents lies at the heart of democracy because political participation simultaneously directs government action, by influencing elections and policy, and confers legitimacy to the government, by assembling support among the electorate for policy positions.

In many modern-day democracies, institutions that support development of pluralistic political consensus feel frayed at best. Political conversation now largely occurs in digital spaces that disincentivize shared understanding and incite demonization by monetizing outrage. Political leaders can now take power by feeding off that vitriol. As a result, trust in democracy and governing institutions—the legitimacy conferred by constituents—is decaying globally, and political polarization is rising.[1]

Yet political divides can be bridged. Some humans excel at encouraging meaningful discussions in small group settings. Conversations with skilled moderators can be gratifyingly productive. They ensure that all voices are heard. They block hateful and off-topic comments. Skilled facilitators go even further, actively encouraging participation and fostering deliberation. They highlight areas of agreement and disagreement and help groups reach consensus, or at least shared understanding. This is particularly hard to do in larger groups, but is more achievable with digital technologies, like feedback surveys and teleconferencing.

Political participation is an aspect of democracy that is fundamentally for and of humans. We don't care if machines express agreement with government policy and, if they did, their having done so would confer the government no legitimacy. But we may need AI moderators to help sustain democracy in increasingly complex societies, because there aren't enough human moderators and facilitators for every issue and jurisdiction. AI's superhuman speed, scale, and patience offer moderation and facilitation capability to every group, and encourage more civil, productive, and inclusive deliberative conversation. It can turn a contentious, polarized, hyperpartisan process into a collaborative, productive one.

When researchers at Google's DeepMind tested an AI mediator with a group of 5,000 British participants in a virtual citizen assembly to discuss contentious issues like Brexit and immigration, they found that the AI matched or exceeded human mediators in achieving consensus while incorporating minority viewpoints.[2] This capability can be especially useful at the local level, such as for town planning, participatory budgeting, and school board meetings.

Content moderation is one of the most well-established use cases for AI. While Facebook's policies have been volatile and it is far from an exemplary actor, it has long used a tiered content moderation system, relying on automated and AI tools to perform the first stage of review.[3] Content moderation gives rise to important questions pertaining to suppression of speech, bias, and error, but it is generally a positive use of AI. We all want online platforms to minimize abusive, threatening, and illicit content. If we can eliminate the most abhorrent content by tasking a machine to evaluate it instead of inflicting it on a human reviewer, that is clearly an improvement. Furthermore, no system staffed by humans alone can affordably process the vast amount of content these platforms need to scan.

AI can go beyond moderating discussion to actively promoting consensus. An early example of this sort of system is Pol.is, an online platform launched in 2012 that allows users to submit short statements articulating their views, then vote on the statements of others. Pol.is uses AI techniques to analyze and produce visualizations of the distribution of different beliefs, showing clusters of agreement and areas of disagreement. It's a powerful system, and has been integrated into the Taiwanese political process[4] and used in Kentucky to deliberate on a twenty-five-year plan for the city of Bowling Green.[5] Taiwan has also leveraged an AI platform called Talk to the City to automatically generate summaries of deliberations among hundreds of participants.[6] Technologies like this can be highly scalable systems of deliberative thought and collective intelligence.

Fascinating work is being done to augment systems like these with generative AI capabilities. Unanimous AI tries to blend the effectiveness of small group deliberations with the scale of large decision-making bodies. Its trick is to embed a single AI into each small discussion group, with the assignment of seamlessly moving ideas between groups.[7] Another tool, Cortico, uses AI to synthesize and report outcomes from human discussions, such as convenings they have led to inform journalists about the top issues in the 2024 US presidential election.[8] The Finnish capital Helsinki has engaged about a third of the city's youth to offer feedback on the city budget, using AI to synthesize more than 70,000 responses.[9]

The deliberation-enhancing capabilities of AI may have the greatest traction in multiparty, coalitional political systems that require compromise to form governments. Case in point: Deliberative systems like citizen's assemblies have long been more popular in Europe than in the US.[10] AI tools that scale deliberation will help to form stable coalitions and to integrate multiple citizen perspectives. However, where these deliberative tools may have the greatest potential for impact is in the most highly polarized political systems, where constituents identify with their political in-group on an emotional level independent of their ideology. Political scientists call this "affective polarization," and research shows that deliberation—particularly facilitated deliberation—has the largest depolarizing effect in this setting.[11] Given that the US is experiencing a historic crescendo of affective polarization,[12] scalable facilitated deliberation seems urgently relevant.

In recent decades, the economic and social interactions of people around the world have migrated to digital spaces in ways that heretofore seemed unimaginable. Acceptance of AI moderation and facilitation in virtual spaces will no doubt evolve correspondingly.

In addition to helping people and groups come to agreement, AI will have a place in all forms of organization and persuasion. Everything we have written about AI and political campaigns applies to political movements and negotiations more generally.

For example, labor unions that fight for fair contracts and other benefits for their members face challenges in organizing, mobilizing, and persuading people. Some already use AI to help them in their work. Since at least 2016, unions have used AI chatbots as an organizing tool, helping to disseminate information about company policies, contracts, and labor laws.[13] One Australian union used a chatbot to mobilize individual action on wage theft by answering workers' specific questions about what work they should get paid for, and when then they should get paid for it.[14] The Workers Lab has built AI-powered apps to help employees report workplace safety violations by auto-populating the details of complaint forms.[15]

Using AI for labor organizing faces an underlying asymmetry: Both workers and management can avail themselves of the technology, but management possesses more inherent power. Whereas organizers can employ AI to help connect and empower workers, bosses deploy it to surveil, command, and control workers, as well as to replace them.[16]

This asymmetry is mirrored across many political movements and campaigns. Establishment forces may seek to use AI to dominate individuals and groups with little power and no authority. Out-groups might respond by resisting the exploitation of AI by those in power, but should not dismiss the possibility of using AI—in responsible, ethical ways—to enhance their own power and solidarity. When AI can legitimately substitute for human labor, it will be difficult for workers to sustain a prohibition on automation; but they can capture some of the rewards of automation and not permit them to flow entirely to employers. For example, in 2023, the Writers Guild of America won concessions that simultaneously guaranteed against the displacement of human authors by AI while ensuring their members' rights to use AI in their work as they see fit.[17]

Of course, sometimes finding consensus is the last thing a group wants, because it erodes established power. If your party is firmly in control of government, you may have no interest in compromise. If you are protecting a private interest, you may prefer to exclude others from decision-making. And if you're in a crisis, you might prefer an executive capable of taking unilateral action to any consensus-building process. Despite these incentives stifling deliberation, AI will provide useful new tools for the work of community organizing, for building power in vulnerable constituencies, and for bridging divides. We hope that a populace increasingly weary of political polarization and gridlock may be willing to give it a try.

37

HELPING YOU ADVOCATE

Many of those who care about political issues do not act on that interest. Most people don't have the time to put their point of view in writing or attend a public hearing, and don't have the means to hire a lobbyist to research and advocate on their behalf. AI can help all of us express ourselves more accurately, more articulately, and more frequently on the issues we care about.

Generative AI can transform your policy preference into an email to your representative, or into written legislative testimony. It can generate multiple versions and let you pick the one that suits you; it can refine it iteratively by querying you about the details. It can take this testimony and submit it whenever relevant bills or regulations come up for a public hearing. It could turn your testimony into a letter to the local newspaper, a brief statement of your thoughts for a post on social media, and persuasive text messages to send to your contacts. An AI assistant could periodically consult you about new policy considerations and add detail to your preferences over time, continuously improving the fidelity of your communications to your representatives. AI-assisted advocacy could give people a greater impact on the political process than they have today.

A good example of these capabilities and integrations is the US-based platform called Resistbot. Since 2017, about ten million Americans have sent text messages to this chatbot in order to receive assistance in turning their political sentiments into messages to their elected leaders.[1] A new feature added in 2023 uses generative AI to automatically craft messages on behalf of users. Resistbot prompts the user to share a news article about an issue and a sentence or two about their perspective, then generates a letter for the user to review. Users can also text feedback to the AI, which will then revise their message.[2] These kinds of AI advocacy assistants have already been widely adopted. Researchers estimated that by 2024, about one in five consumer complaints to the US Consumer Financial Protection Bureau was written with the assistance of AI, with usage sharply increasing after the release of ChatGPT.[3]

Expanded use of similar tools will have both good and bad effects. On the one hand, we want politically engaged citizens to better express their views. AI is an assistive technology that can help accomplish that. For people with disabilities, this capability can be especially powerful. In 2024, Jennifer Wexton, a US representative from Virginia, delivered a floor speech using AI after losing her voice to progressive supranuclear palsy, a neurodegenerative disease.[4] Decreasing the effort required for correspondence could help a greater diversity of voices be heard by policymakers.

On the other hand, the signaling function of these communications changes when they can be produced with less human effort. Today, when you take the time to write a letter to your representative, that action often says more than the opinion you express. It says that you feel politically empowered and committed to having your voice heard, and that you should be heeded because you are both a voter and capable of influencing other voters. It also says that the topic is something about which you care deeply: Either the issue has ramifications for your life, or you are morally affected by it. If AI becomes widely used for these communications, both of these signals disappear.

If lawmakers receive comments from both people who care deeply about an issue and others who are just piling on, how will they decide which issues and opinions to prioritize? This is hardly a new phenomenon. For decades, advocacy organizations have asked their members to participate in coordinated lobbying campaigns, and have provided form letters to enable them to deluge policymakers with their opinions. Social media has magnified this capability; millions of individuals and bots now steer and stoke public discourse on any given issue. The popular wisdom used to be that if one person complained, ten others just as angry didn't have time to complain. Now, if ten people complain, there might only be one angry person.

This enhanced communication capability leads to a phenomenon called "consensus by attrition,"[5] common to local governance bodies such as school boards, zoning boards, and town councils. The loudest, most annoying, most stubborn person often gets their way by simply wearing down everyone else in the room. AI helpers could make it easier for anyone to be that noisy person.

Sometimes we want it to make it slightly difficult to initiate a bureaucratic action, like registering a complaint. If it's trivially easy and cost- and risk-free to challenge a ruling you don't like, then everyone would do it all the time. The automated appeal would become a normal part of the system. AI advocates will be able to tirelessly argue and petition on behalf of and against even the most stubborn human. Systems of governance, like public comment systems and appeals processes, will have to adapt to this new reality.

In general, we favor assistive technologies that broadly help people access the political system. One of our projects, the Massachusetts Platform for Legislative Engagement, uses both traditional web technologies and AI to help individuals advocate to lawmakers.[6] We are excited by opportunities that promote individual expression across barriers of language, articulation, and free time to help people more fully participate in the democratic process. We prefer the problem posed by legislators having more input than they know how to handle, to the problem posed by effectively limiting political participation to only the most privileged few.

AIs employed by legislators and citizen advocacy AIs could eventually find themselves in an arms race. The former are trying to discern what people think, and the latter are trying to have their individual voices recognized. We might have a lot of AIs talking to other AIs, but that wouldn't necessarily be a failure. It would not be so different from the intermediation we have now, with citizens represented by interest groups, who speak to staff, who represent legislators. Mediators are a natural component of the information system of democracy.

Beyond helping people express their views, AI can help people interact with government agencies. AI can already assist people to navigate such tasks as filling out forms, applying for services, or contesting bureaucratic actions. This will make it easier for people to get the information they need and receive the benefits to which they are entitled.

A simple but effective example of this type of assistance is Google's "Hold for Me" feature, which listens when you are on hold with a call center, and prompts you to speak when someone finally picks up the phone. The amount of trust required to adopt a system like this is low; it only needs to be seen to function properly one time to convince you that it is worthwhile.

The trust barrier for other bureaucratic support use cases, like ones that require access to your personal financial or health information, are much higher. One early study conducted in England by the AI company Limbic created a chatbot that assisted patients with access to mental health resources, and found a nearly threefold increase in referrals for those who used the chatbot over those who did not.[7] More importantly, the chatbot significantly increased referrals among minority populations, including Asians, people of color, and especially nonbinary individuals.

These capabilities will grow in sophistication. The AIs themselves will improve, and they will be better integrated with online services, both government and commercial systems. An AI that knows how to fill out a form is useful. An AI that, securely and under your control, can access relevant data from your city's website, your email history, or your phone when filling out that form would be even more so.

While you can hire people to help you navigate byzantine government processes, the people who need them most generally can't afford one-to-one assistance. AI is a more affordable option. mRelief, a nonprofit startup in the US, is building AI chatbots to help people apply for nutrition benefits.[8] The US nonprofit Justicia Lab created IMMPATH, an AI tool that provides personalized guidance on immigration proceedings. Users upload photos of immigration agency correspondence, then an AI assistant explains how to respond and provides connections to local immigration advocates.[9] This need for AI assistance would not exist if the relevant agencies were adequately staffed to help the communities they serve, and will become increasingly important if the US government continues to reduce its workforce.

AI systems that are as trustworthy as human helpers would seem to be an unmitigated good. However, not everyone wants accessing government benefits and services to become easier. This difficulty in dealing with government systems is known as "administrative burden," and sometimes it is deliberately introduced.[10] If a legislator opposes a benefits program that is nonetheless enacted, they can reduce its effectiveness by intentionally making it difficult for people to sign up for it. NGOs and AI tools generally don't need to ask anyone's permission to help ease administrative barriers, although, as with fiscal and healthcare proxies, governments may pass specific restrictions on AI proxies used for this purpose.

We can sympathize with those who might never choose to expose personal information to an AI, or to rely on it to help in sensitive contexts, or to accept that their government would defer its service responsibilities to machines. Some resisters will be won over when a concrete opportunity to save time or money presents itself. Others will, rightly, question whether they can ever trust AI tools to act in their interest. Many of the beneficial use cases in this book will never be implemented until we can make AI truly trustworthy.

38

ACTING AS YOUR PERSONAL POLITICAL PROXY

Representative democracy requires elected officials to stand in for the collective preferences of their constituents. The most understandable reason for this is logistical: Citizens don't all have the time or ability to communicate our preferences directly, and we can't all fit into the legislative chamber to debate the issues at play.

This system doesn't always work very well. We elect representatives and send them coarse (sometimes downright obtuse) signals about what we want, then rely on them to extrapolate the details. Elections tend to be framed around a few key issues—immigration, taxes, healthcare—with each candidate lined up on one side of an issue or the other. Whoever wins assumes a mandate to figure out the rest. Sometimes we give a little extra signal, such as up or down votes on a few ballot initiatives. Our elected officials do talk with some of us in person, usually a few at a time, and get bulk signals from polls and social media. The result is that not much information is exchanged between people and their representatives, which can make it hard for the public to ensure their representatives' actions align to constituents' preferences.

AI can change this. As we discussed in the previous chapters, AI could ascertain our views, extrapolate our political preferences, and advocate on our behalf. This might happen first in nongovernmental contexts. AI proxies have been proposed to vote on behalf of investment funds that control corporate shareholder votes.[1] For citizens, an AI personal assistant trained on and continuously attuned to your political preferences could advise you on where to make political donations, which candidates to vote for, and which ballot initiatives to support. Already there are websites that do this after asking you a few questions; an AI personal assistant could be a much more sophisticated virtual guide.

Once AIs are commonly in use, they could shift the balance of power between lawmakers and their constituents. In particular, they could make ballot initiatives—direct democracy—a more effective and efficient means of enacting policy, which wrests policymaking from the hands of legislators.

Today, most ballot measures are strictly binary: A proposal is precisely defined, and citizens can vote yea or nay. With such a system, even popular policy ideas can fail if the wording of the proposal is slightly off. Proponents of a failed initiative might have to wait years for the next ballot cycle to try again. This limitation has made direct democracy far less flexible than legislating conducted by elected representatives, who have the authority to negotiate, trade, compromise, and refine a bill, and to call votes again and again until they arrive at the finish line.

Individual voters lack these options because there are so many of us, and because we're busy. We don't have time to vote every day and we can't spend our days researching dozens of issues, or dissecting the language of each of the bills considered by our legislatures.

An AI proxy could develop a good understanding of your policy preferences and could help you decide how you will vote. If your position on a new ballot question is clearly indicated by your past behavior, such as your voting history and past interactions with the AI, the AI wouldn't even need to consult you. If the AI cannot confidently determine your position, it can ask you clarifying questions. If you trust your AI proxy to provide relevant background information, you might not need to conduct independent research. If AI tools like this are built well, and are embraced by both citizens and officials, government could become more responsive to popular input and get more done. That's an awful lot of "ifs" and "coulds" about trust, but it's a natural extrapolation of the advocacy tools we talked about previously.

The most obvious way AI can improve this situation is to help us respond to an increased number of ballot initiatives. Today, nonpartisan voting guides that present the case for and against each initiative on a ballot are a valuable tool for helping voters decide on these questions. These guides often feature several dense pages per question, so they don't scale well for human readers. If AI could make it easier for you to decide how to vote, people could vote on far more ballot initiatives. Future ballots could have hundreds of items, and be voted on more frequently, allowing voters to control policy with a level of specificity previously unheard of.

Far more speculatively, voters could authorize an AI proxy to vote on their behalf. If voters were individually represented by efficient, effective, trustworthy AI proxies, they could be consulted—through their proxies—more frequently and in greater depth. Ballot initiatives would not need to be binary, or to align with election cycles. Through their proxies, voters could be polled continuously and with an elaborate menu of options, thereby capturing a far more nuanced representation of our individual and collective preferences.

Even more radically, AI could facilitate a new kind of direct democracy, where a combination of voters and their AI proxies collectively make policy decisions. We can imagine a political system where a personal AI is trained on your wants, needs, beliefs, ethics, and political preferences, and then becomes your personal representative in a massive legislature, where millions of these proxies collectively debate and then vote on legislation. If issues you care about most are on the table, your AI assistant could ask you to submit your opinions and votes directly.

This is certainly not democracy as we know it today, but researchers are creating AI tools that could someday enable these new forms of governance. Computer scientists are currently developing a concept called generative social choice, whereby AIs learn individuals' policy preferences and, collectively and iteratively, generate policy proposals to which a large majority of the humans involved would demonstrably agree to.[2] Eventually, systems like this could even propose legislative language.

AI proxies could help represent the interests of nonvoters in the legislative process: children, animals, even future generations. We could instruct AI proxies to appropriately represent the interests of these groups.[3] This doesn't equate to extending the franchise to particular AI personas, but it could mean that deliberation and policy formulation could include AI representatives that favor policies that support, for example, forest preservation.

This may sound like fantasy to the uninitiated, but the challenge of internalizing entities who are not franchise holders in a political system is a well-studied problem in economics, law, and political science. UN conventions call for representation of children in legal proceedings.[4] Hundreds of jurisdictions recognize "rights of nature," analogous to human rights.[5] Choosing to develop and integrate AI systems that speak directly on behalf of these out-groups may be one way to accomplish that. Going further, if the AI systems currently being developed to communicate with animals make meaningful progress, established ideas of non-human representation take on new significance.[6]

Of course, implementation of AI proxies can go badly wrong. First, as we've discussed in so many cases, this would require tremendous trust between voters and their AI proxy. This won't work if you worry that it might misunderstand your intent, or misrepresent a ballot question. It won't work if proxies are inherently biased towards someone else's views, such as the views of the corporation that built it. And it won't work if you have to worry about your AI getting hacked.

AI proxies might not provide an accurate representation of your political views. An AI that extrapolates from one response to another might be wrong. This is a classic problem in both statistics and AI; there are methods to understand when a predictive system fails to generalize from one domain, such as one type of ballot question, to another. An AI proxy can guess at your response to a new policy question—real or simulated—and check with you on the result. If the proxy gets it wrong, you can tell it so and thereby update both the tool's understanding of how often it needs to consult you and your own understanding of how reliable it is. Everyone will have a different tolerance for this kind of failure, so the AI proxy could be trained to consult with its owner as often as they feel is appropriate.

Policy uncertainty also happens in traditional representative democracy, but it is generally resolved rather slowly. Laws are repealed and replaced, amended, challenged in courts, and updated ad nauseam for years. Sometimes policy changes are so chaotic that their implementation is stalled while the legislature, the courts, or the voters sort out what they really mean. This often indicates an unstable equilibrium, a policy formulation that is vulnerable to easy toppling by pressure from any direction. With better representation of our collective preferences, we might find more stable policy formulations. All of society would benefit if that AI could help lead to policies that were truly better for the people they affect and required less political struggle to implement and maintain.

The most optimistic reason to develop systems like these is to make government more equitable. One problem with representative democracy is that it can be anti-majoritarian. We citizens don't each get our own representative; entire communities vote to select one person who's supposed to represent all of us. If your democracy does not employ proportional representation and you're part of a demographic that is consistently in the minority across districts (possibly due to gerrymandering), you might find yourself persistently unrepresented. Just as bad, many representative bodies, such as the US Senate and the Japanese National Diet, do not apportion representation according to population, so some groups have even less power than their numbers might suggest. Even direct democracy can be inequitable in practice, because voter participation is often lower among less affluent or less motivated groups. If everyone has equal access to an AI proxy, those barriers to participation could be lowered if not eliminated, and everyone could have a more equal say.

More generally, instead of our democratic systems of preference aggregation being constrained by humans' capacity for complexity, it would be constrained by AI's much greater capacity for complexity.

How alien would the world be if AIs were casting ballots and making decisions for us on a grand scale? We claim not very. Every denizen of a complex society has a lifetime of experience ceding the most important decisions governing their lives to a faceless machine.

Most of us want to believe that we can intervene when it matters. If an institution acts aberrantly, corruptly, or harmfully, we can apply political pressure to force a change in its behavior. We can vote out the elected officials accountable for them. We can use a ballot measure to force a policy change. We can picket outside the institution's headquarters. This collective individual expression will bring about change. If not, democracy isn't working.

The most foundational concern pertaining to the implementation of AI proxies is that it could contribute to the atrophy of our innate capability for democracy. Aristotle envisioned democracy as a form of self-actualization. In this view, individual political preferences don't magically appear in our heads, just waiting for an AI to discern them and then to advocate on our behalf. They are created through the process of education, discussion, and debate. The act of doing democracy makes democracy work. If we allow our AI proxies to fight it out then tell us the result, we lose out on that human interaction. Moreover, if we begin to listen and believe the AI's version of our preferences, we could lose any sense of personal conviction. This could easily result in the same sort of feedback loops that propel people towards extreme views, or to entrench and polarize political parties. Replacing humans with AIs could result in worse outcomes in the long term, even if it led to better policy decisions immediately.

Of course, the outcome depends on the application. We might be okay with AIs driving our cars, and with our forgetting how to drive—just as few of us know how to drive a horse and buggy. We're probably not okay with AI passing all laws while we humans forget how to do democracy. Maybe we will allow the AIs to help us deliberate and build consensus, but nothing more.

The actual decisions made by a government, even in the policy areas about which we care viscerally, are as remote to most of us as are the biochemical reactions in our cells. Parents care deeply about their children's education, but most don't have the time to read the texts assigned at public schools—much less to exercise influence over which texts those are. Citizens who care deeply about fairness and justice generally don't have the faintest idea what cases are being tried inside their local county courthouse, much less what rules are being applied or the qualifications of the prosecutors and judges acting in their name. It's simply impossible for citizens in a modern democracy to be informed of or to formulate opinions on most of the ways government functions. For generations up to the present day, the solution has been to entrust that authority to institutions. We even trust those institutions to choose technologies that help automate their work.

AI is just another step down that same path: another technological option for us to use in expressing our policy preferences and for institutions to use in executing their functions. If we can trust the AI, it might even be a positive step.

Most of these scenarios are too far into the future to predict clearly. What we want, right now, is to enhance human civic engagement, even as AI becomes essential as an assistive technology. In the next and final part of this book, we'll explore how we can maximize AI's benefits for democracy while minimizing its risks.

VII

ENSURING THAT AI BENEFITS DEMOCRACY

By now, we hope you are convinced that AI will have a real impact on democracy and that, in spite of the risks and challenges that exist, it's possible for that impact to be positive. We've tried to illustrate the many and varied applications of AI, and how exciting and scary they may seem. However, we haven't yet laid out what should be done about it all. If our goal is to ensure that AI generally benefits democracy rather than harms it, then we have a lot of work to do.

39

WHY AI THREATENS DEMOCRACY

In this book, we've outlined a variety of ways that AI can change democracy, both for good and for bad. We've discussed both immediate and longer-term effects, and have highlighted where things can go right or wrong. We've tried to keep an open mind to the potential to apply AI as a force for a more inclusive, more equitable, and more functional democracy, but nonetheless recognize that this is not the path being taken by many governments.

This feels like a particularly precarious moment for democracy as an institution. Democracies around the world are experiencing a rise in authoritarianism paired with erosion of their traditional institutions (consider Hungary, Türkiye, Russia, and—increasingly—the US). The interplay of destabilizing factors like climate change, an upturned world order, global migration, and widespread corruption is driving economic instability across both affluent and developing nations (consider the extremes of Venezuela, Sri Lanka, Kenya, and the US). Many democracies are experiencing a surge in political polarization that makes governance feel unstable (consider France, Brazil, Germany—and the US). A changing international order is further stressing democracies around the world. And that's before injecting a transformative new technology that could disrupt everything.

There are many ways in which AI could accelerate these disruptive trends. As we introduced in Part I and have reiterated throughout this book, it could exacerbate injustice, fail to demonstrate trustworthiness required for sensitive democratic applications, and intensify the concentration of power rather than equitably distributing it. Our future might include opportunistic political candidates leveraging AI to disseminate false information, lobbyists deploying AI to enrich their elite clients, and overworked, AI-augmented public defense attorneys forced to take on caseloads that are still larger than they can handle. AI could empower demagogues to cater to the worst proclivities of voters at an individual level. It could afford governments new means of spying on their citizens, curtailing their liberties, and otherwise abusing power. And, as quaint as this concern increasingly seems, AI deepfakes could result in election chaos.

AI is a technology that can help the rich get richer, and the powerful become even more powerful. We've already seen this begin to occur. One of the most glaring concentrations of power produced by AI benefits its corporate developers. The companies are growing at a staggering pace by promoting AI. We have written about how both states and corporations are long-standing forms of superhuman entities, what science fiction writer Charlie Stross called "slow AI." The difference is that democratic states are obligated to act on behalf of the interests of the whole public, while corporations are constructed to serve some private interest.

Big Tech companies already have scale and resources surpassing all but the largest nation-states and are pursuing

trillion-dollar profit potential in AI. Just as streaming platforms have gobbled up media industry revenues and rideshare platforms have subsumed ground transportation
markets, AI developers believe their products will eat into
the profits of human-driven services that involve cognition,
like education, journalism, and lawyering. And in the spirit
of "disruption," they may enter these markets with disregard
for historical ethical norms, best practices, and standards
that have been developed in each field.

The power of Big Tech has engendered a new class of
oligarchs that now wields political, cultural, and financial
power on a scale unprecedented in history. Big Tech leaders
are clearly willing to spend those revenues to protect their
market positions, as demonstrated by their massive lobbying investments in the US, EU, and elsewhere.[1] A feedback
loop has emerged whereby investment in AI drives lobbying
spending, which clears the regulatory field for further investment in AI.

This lack of AI regulation and concentration of power has
already exacerbated injustice and caused harm. AI development is exploiting people, as well as mineral, energy, and
water resources, particularly in the Global South. Social
media has been broadly seen as stoking polarization, sometimes to the point of promoting political violence.

The most important questions raised about AI pertain to
choices. The forces of technology, powerful private interests,
and a nexus between these elements and democracy have
already been unleashed. What remains is to decide what to
do about it.

Good reasons exist for people not to trust today's AI. AI
models that hallucinate generate concerns about accuracy,

and models that can't provide explanations generate concern about bias and other unfairness. AI systems that are insecure are widely and accurately recognized as unsafe to use.

However, we're past the point of wondering whether AI could or will be used in politics and government. Around the world, politicians, political parties, advocacy groups, and courts already use AI tools to assist them in their work. Entities that find efficiency and advantage in using AI will spur their peers and competitors to do the same. Corporations began marketing AI tools to bureaucracies, law enforcement agencies, and other parts of government as soon as they were even slightly effective, and well before they were shown to be fair, equitable, and unbiased. The latter still hasn't happened.

The fundamental threat of AI to democracy does not lie in its technology or capabilities, but rather in the fact that it is being developed and used by the powerful to further their own interests. Consequently, it is not yet trustworthy to use in democratic contexts. To solve this problem, and make AI work for rather than against democracy, we need to find ways to make AI more trustworthy and to distribute control of the technology. That requires thinking creatively about how to make AI more democratic.

40

HOW TO BUILD AI FOR DEMOCRACY

Technology will not solve democracy's problems, but it already has a powerful influence over democracy. We want to shape the emerging technology of AI into a force that can promote good democratic governance, to be used with the best interests of democracy in mind.

This will require that governments grapple with AI, and not simply consign its development and application to corporate entities. Regulation is a key tool for governments to impel the development of AI that we can trust to support democracy. We need structures that incentivize, and laws that enforce, the trustworthiness of AI. We need transparency into how AI systems are built, when they are used, and what biases they encode in democratic processes. We need AI—and robotic—safety regulations that encompass all of the ways in which AI affects the physical world and alters the speed, scale, scope, and sophistication of human activities.

The EU has taken the most comprehensive and boldest action on AI to date, by means of the 2024 AI Act, and deserves credit for its early regulation of this new technology. For example, the act requires disclosure to employees of certain AI uses in the workplace and prohibits AI from exploiting and manipulating people with disabilities. However, the AI Act falls short in important ways. Although it creates a public registry for disclosure of high-risk AI systems, fairly narrow boundaries define the sort of models that must be disclosed. Even where it requires disclosure of AI implementation, it does not require AI developers to respond to public input, or solicit any engagement by stakeholders. For example, while AI developers are required to have a mechanism to lodge complaints for rightsholders who suspect copyright infringement in AI training materials, it is unclear if AI companies have any obligation to reply.[1] Also, national security applications are exempted from many of its safeguards, meaning that the most sensitive possible uses of AI in democratic contexts are subject to the least regulation.[2]

The US Congress, meanwhile, appears too gridlocked to undertake any meaningful legislation on AI, and the Trump administration has vocally decried AI regulation. The UK has stepped back from regulation as well,[3] a shift the EU is being pressured to follow.[4]

To be effective, regulators must shift their focus from the technological minutiae (e.g., how big the models are, whether the code is open source, what capabilities a model has) to the real source of risk in the world of AI today: the choices made by people and organizations. Remember that AIs are not people; they don't have agency, beyond that with which we choose to endow them. Recall the discussion in chapter 14, "Who's to Blame?" AIs are built by, trained by, and controlled by humans, who are mostly employed by for-profit corporations. To be effective, AI regulations must meaningfully constrain the behavior of corporate AI developers. Regulations on the behavior of AI users—including political campaigners, police, attorneys, and others—should also be reformed, to confront the enhanced speed, scale, scope, and sophistication conferred by AI. Penalties for violations must be large enough to deter lawbreaking, and there must be resources for enforcement.

There is much to be done in addition to AI regulation. Every actor in the AI ecosystem has a role to play, and can independently take positive action. Any one of the leading AI companies could voluntarily allow public input to help steer its own work. Industry associations could adopt standards and ethical norms for AI development, and mandate systems of oversight and penalties for misconduct. Researchers could continue to aggressively oversee the activities of industry and its products, resisting pressure from political factions opposed to government oversight. Governments could leverage their buying power, restricting procurement from vendors who do not adhere to trustworthy AI principles. And, although this isn't a problem that should fall on individuals, all of us can shift our own spending towards those companies who demonstrate good practices.

It would be good to have voluntary, system-wide commitments to making AI work better for society, and many organizations have demonstrated their willingness to work with these types of approaches. However, tech companies have shown us that they will fight hard against restrictions on their ability to profit from AI. One case in point is the massive 2024 industry lobbying campaign that led California Governor Gavin Newsom to veto a popular bill imposing basic safeguards on AI development, like whistleblower protections for AI engineers and safety planning requirements for large companies.[5] We should not expect anyone—particularly corporations—to act against their own best interests. We need to apply all the systems of democracy to orchestrate multilateral governance of AI.

The more AI is adopted by individuals and institutions, the more the concentration of its control threatens democracy. We need to find ways to distribute control of this technology. This does not mean eliminating corporate AI; rather, it requires robust noncorporate alternatives. Ultimately, AI that works for democracy needs to actually be democratic, and that requires something more than today's corporate AI systems.

One of the most exciting opportunities we see to distribute control over AI and improve its trustworthiness is to spur development of AI from noncorporate actors, particularly governments. Many national governments have the resources to develop their own AI models, and could provide these free or at cost as public goods. Although the cost of training the most sophisticated AI models runs in the hundred-million-dollar range, most governments need not invest at that level. Governments do not need to compete to be at the cutting edge of general purpose model performance; they could leverage open-source starting points, and benefit from widespread innovations to drive those costs downward to a few million dollars, or less. Governments like Singapore[6] and Indonesia[7] are actively building public, open AI following this approach; they take publicly available building blocks and localize them for language and use with government applications. Similar pilot projects are also being conducted in other countries, like the US[8] and France.[9]

The compelling difference between public and private AI development lies in incentives. In a democracy, government entities have an obligation to serve the public, and need not be directed by the profit motive. An emerging category of products called Public AI, AI developed under democratic control and for public benefit, can more easily adhere to democratic principles because its constituents require it and because its developers have no shareholders to placate financially.

This kind of AI could represent a new, universal economic infrastructure for the twenty-first century, much like the public schools and highway systems built by past generations. Democratic control is a fundamentally different basis for trust than corporate AI, but it's no panacea.

Public AI could be nationalistic, built by one nation according to its own values and flaws and for the benefit, or oppression, of its own citizens. Americans are not likely to trust Public AI from China, and vice versa. Alternatively, Public AI could be transnational, generated through international cooperation between nations that may vary in their resources, but that share values and goals. The US and Japan are already collaborating to build a trillion-parameter AI model focused on scientific applications.[10] Even when there is no formal agreement between states, this kind of international engagement may be inevitable. The 2025 release of open-source models from the Chinese company DeepSeek motivated massive interest and experimentation with their tools by AI developers around the world, including efforts to remove the models' evident censorship and bias.[11]

You may not trust your own government to control the development of AI. You may not trust the AI developed by another government. But democracies give us more say in how they operate than corporations do, especially in non-competitive markets. Governments can fulfill use cases for trustworthy AI that industrial players cannot. The goal is for Public AI to coexist with private AI, providing a bulwark against monopoly and setting a competitive baseline for transparency, availability, and responsiveness that commercial competitors need to meet in order to succeed.

Moreover, Public AI models would have a more justified place in democratic contexts than corporate AI models. Democratic governments should not employ secret and proprietary technologies in making civic decisions and taking civic actions; they need an alternative to corporate systems when using AI to assist in and automate functions of democratic decision-making.

These alternatives need not be limited to national governments. They can be developed at the state and local level.[12] NGOs and nonprofits can also build AI models, although the substantial funding required for AI development makes them highly vulnerable to corporate influence. The saga of OpenAI, founded with a public-interested mission before it took billions of investment dollars from Microsoft and abandoned its nonprofit model,[13] demonstrates that private entities can be tempted by profit to undermine their values and reverse their public commitments. Public entities in a democracy can be corrupted too, of course, but they at least offer electoral means of public control.

The market also has a role to play in shaping corporate behavior. Every technology company seeks to differentiate itself from competitors, particularly in an era in which yesterday's groundbreaking AI quickly becomes a commodity that will run on any kid's phone. As a leaked Google memo famously acknowledged in 2023, technology companies have no sustainable proprietary advantage when it comes to the latest and greatest technical achievements of their models.[14] Yet they can differentiate themselves by building trust with their consumers. This will be a challenge, particularly because so many tech companies and their leaders have done so much to burn trust over the past few decades. But it's possible.

We urge corporations, and particularly their AI developers, to treat public trust as a valued asset. That means thinking holistically about the transformations they are spurring, exercising greater transparency in AI development, and being meaningfully responsive to public input. The difficulty in achieving trust makes it a particularly valuable asset. AI will only inspire trust if its developers make sincere, consistent, and verifiable commitments to consumers in the areas of transparency, privacy, reliability, and security.

It is foolhardy for tech companies to sacrifice democratic values for short-term gain. Plutocrats lobbying for favorable treatment from autocrats abroad, or cozying up to authoritarian movements in their own countries, should look to the deposed oligarchs of the twenty-first century to see what can happen if you build wealth on the beneficence of autocrats. Chinese decabillionaire Jack Ma's wealth dropped by half in 2020 after he criticized the Chinese government.[15] The then-richest man in Russia,[16] industrialist and Kremlin critic Mikhail Khodorkovsky, was imprisoned and then exiled by Vladimir Putin.[17]

Enabling democracy is good business and undermining it is an existential risk. The opportunities we have described in this book to use AI to assist campaigners, government agencies, courts, and other functionaries of democracy are potentially highly profitable. Corporations will limit both their own and society's potential if they build untrustworthy AI that engenders corruption, stokes inequity, and exacerbates oppression.

41

PRINCIPLES FOR AI THAT HELPS DEMOCRACY

To build AI that works for democracy and that can bring about more benefit than harm, we need organizing principles under which each party involved in developing and operating AI can operate. This would represent a shared vision of democratic AI. We propose the following:

- AI must be **broadly capable**. It needs to satisfy the full range of AI use cases: from predicting outcomes and making decisions, to generating text and images, to still other applications that will become possible in the future. How that works may change over time. It could mean a single model with broad capabilities, or it could mean a handful of specialized models with narrower capabilities. Although they don't need to be the world's best-performing AIs, they do need to be capable of executing each job adequately.

- AI tools must be made **widely available**. Anyone, including public and private entities, needs to be able to access them, regardless of their identity, affiliation, nationality, or wealth. They must be unrestrictive in their licensing, portable to many different hardware platforms, and available for use at low or no cost.

- AI developers and tools must be **transparent**. In a democracy, the consent of the governed requires an informed citizenry. Systems and technologies of governance must therefore be open to review and criticism. AI tools must be available for study, testing, and extension, requiring models to be open source with respect to both the code and the data.

- AI developers must be **meaningfully responsive**. The public must be invited to participate in AI development and management. Entities developing AI—whether a for-profit corporation, a nonprofit, a nation, or an international organization—should actively solicit and demonstrably respond to input from stakeholders, and developers should be held accountable if they do not do so. Responsiveness does not stop with choosing data inputs and model designs; AI developers must consider and monitor the winners and losers from any deployment of their products, and to responsively mitigate disparate impacts and potential harms to different communities.

- AI must be **actively debiased**. As we've discussed, there is no one true definition of fairness, equity, or bias, and the last thing we want is for AI to step in to decide these questions. We must recognize that human choices drive the biases of AI systems, and acknowledge and manage their effects. No system can please every demographic all of the time, but no basis for trust will exist if AI systems deny or obscure users' values.

- AI tools must be **reasonably secure**. They need to do what they promise and nothing more. Using them must not result in the disclosure of confidential or sensitive information. They must enforce data integrity and reflect the actual state of the world. They must be subject to public direction (see "meaningfully responsive" above), not surreptitiously beholden to private interests such as their developers, their billionaire investors, or antidemocratic actors (like authoritarian states). They won't be infallible, but they must be uncorrupted.

- AI tools and their developers must be **nonexploitive**. They must not co-opt public labor for private profit, whether that public labor consists of millions of individuals' collective creative output used as training data, or workers in developing economies used to annotate images without being fairly compensated. The actual costs to people and environments must be included in any calculation of the net value of an AI system. Corporate AI developers are skilled at externalizing these costs, but AI aimed at enhancing democracy must demonstrably not do so.

These principles for trustworthy AI for use in democratic contexts are the same principles we would insist that any publicly accountable institution adhere to: to be capable, available, transparent, responsive, debiased, secure, and not exploitative.

We are deliberately not prioritizing other issues. AI does not need to be free from mistakes; it can be useful without being perfect. Neither does it need to be free of risk; beneficial uses for AI exist even in fields where it could cause harm, such as biotech. Also, AI does not need to be altruistic; some entities might profit from it. If we saddle AI with too many additional requirements before we consider it suitable for use in democratic contexts, it will never be possible.

Many people and groups advocate for responsible and trustworthy AI. In 2020, Mozilla published a white paper

on trustworthy AI with many of these same ideas.[1] Data &
Society's Michele Gilman[2] and the Ada Lovelace Institute's
Lara Groves[3] have each originated frameworks for public
participation in AI development. The Collective Intelligence
Project has outlined what it calls "Democratic AI," a definition for an AI ecosystem that will "provide public goods, and
safeguard people's freedom, wellbeing, and autonomy"; that
is, a definition of AI that behaves like the ideal of democracy
itself.[4] Government agencies like the US National Institute
of Standards and Technology have put forward their own
guidelines for trustworthy AI.[5] Meanwhile, research insti-
tutes monitor the industry's adherence to ethical principles,
such as the Stanford Center for Research on Foundation
Models' transparency index, which measures factors like
model accessibility and data disclosure.[6]

Widespread implementation of these principles will not
be easy to achieve, and there is no easy solution for any of
the concerns we have discussed. Broad capability largely
exists, but the other factors do not. Corporate AI models
incorporate structural incentives to limit their availability.
For example, OpenAI gave its lead investor, Microsoft, first
dibs on using and commercializing its key technologies.[7]
While the R&D and operational costs of many of today's
leading models have been subsidized by venture capital,[8]
ultimately, corporate AI developers will extract costs from
users by hiking fees or through monetization mechanisms
like advertising. OpenAI, for example, has signaled its intent
to charge exorbitant rates for its most capable tools.[9] Corporate AI developers have demonstrated that they have no obligation, and little inclination, to shoulder the environmental
costs of their products and have shirked their responsibility

to publicly acknowledge and mitigate the disparate, harmful impacts of their products.[10] Most of all, corporate AI has failed at being responsive. Even Meta's proclaimed open model, Llama, is developed largely out of public view; even the data used to train it is kept secret.[11]

The barriers to trustworthy democratic AI are more social and political than technical. For example, steering corporate AI towards these outcomes requires governments to more actively regulate AI developers to avoid exploitation, to eliminate subversion of user interests on behalf of advertisers, and to require developers to meaningfully consider public input. Providing robust alternatives to corporate AI requires political will, and perhaps even significant public investment, amongst many other pressing needs.

42

THE RIGHT WAY TO REWIRE A DEMOCRACY

Reporting on AI's future tends to focus on the possibility of an AI apocalypse or its role in a new Cold War: an arms race for AI supremacy between the US and China. It's no accident that these framings dominate public discourse, because they bombastically aggrandize tech company elites and serve the interests of their billionaire investors. However, AI doomsday and superpower competition scenarios are the least of our concerns.[1] The near-term possibilities of undermining democracy by means of AI are much more realistic, and require our urgent attention.

AI matters to society because it will change how people interact, often to the benefit of the privileged and the detriment of the vulnerable. To protect the interests of all members of a pluralistic society, democracies must look beyond the isolated, technical aspects of AI to recognize that social policy is technology policy.

Like other technologies, AIs are just tools. The laws governing social behavior—civil rights, public conduct, public safety—dictate how those tools are used, even when they are not specific to AI. For example, AI matters to the economy because investors, startups, and employers all see it as a means of squeezing new profits out of old ideas: selling AI-enhanced versions of familiar software, making workers more efficient, and automating cognitive labor that people used to perform. Economic policy is thus also technology policy.

Therefore, solutions need to be responsive to the risks heightened by AI and not specific to its technology. For example, there is no technical means of preventing AI from assisting autocrats, but legal and institutional fixes do exist. Democracies have centuries of experience organizing a non-partisan, professional civil service to compel loyalty to the law and the public interest rather than the elected leaders that hire them. The problem of AI misuse can be overcome by strengthening those civil service protections—and keeping human civil servants in charge of AI.

We need to rewire our democracies to sustain them through a new age of technological transformation. In the near term, we urge action along four paths that are both necessary and demonstrably successful.

- First, **reform** of the AI ecosystem. Governments need to actively steer the development and use of AI to be more ethical, just, and equitable. Reform should involve regulation of both AI and tech industries, and aim to diversify control of AI. Regulation is critical, just as it is for any industry that has such a consequential impact on our political, economic, and societal systems. However, AI development must not be relegated to the exclusive discretion of a few private, for-profit corporations; the speed, scale, scope, and sophistication of AI represents a risk to democracy. Public AI development offers a fundamentally different path forward to create trustworthy models and to fulfill the principles we have outlined for achieving AI that helps democracy. AI development that follows these principles would be safer than present-day AI, and could be one part of a bulwark against authoritarianism.

- Second, **resistance** to harmful uses of AI. AI engineers should resist being assigned to develop unethical applications of AI. Agency staff responsible for administering benefits should oppose handing over that authority to unvetted or poorly performing AI systems. Citizens should resist allowing their governments to use AI to erect a totalitarian surveillance state. This resistance will be hard work, both because of the powerful entities with vested interests in deploying AI and the countless fronts on which AI will be repeatedly introduced.

- Third, **responsible use** of AI in society. There are exciting opportunities right now for beneficially leveraging AI in our democracy—a positive agenda that complements resistance to AI. Speedy machine translation can make linguistically diverse democracies more inclusive, and enable political leaders to engage with more of their constituents. Sophisticated AI tools can help legislators make their laws more precise and complex, and more responsive to constituents' input. The scale of AI can help administrators and courts improve the delivery of government services and eliminate ministerial delays that plague many modern bureaucratic processes. Perhaps most excitingly, the scope of AI can help amplify citizens' voices, and assist them in injecting their perspectives and experiences into policymaking processes that they might not otherwise have the time or knowledge to influence. We need to responsibly experiment with using AI in democratic contexts so that we can usher these positive visions into reality.

- Fourth, **renovation** of society for AI. We need to prepare our democracies for the impacts of AI by adopting reforms that respond to the challenges it will bring. The speed, scale, scope and sophistication of AI will overturn assumptions of current law and democratic processes. It will lower the barrier to entry for filing legal cases and administrative appeals, overwhelm the court system, and help lobbyists execute their strategies with superhuman scale and scope. Some reforms to enhance many of these systems are familiar, such as increasing the capacity of courts, and limiting lobbying activities and campaign donations. Many jurisdictions have been lagging on good governance since long before AI became a pressing concern, and it is past time to act. This is what political scientist Danielle Allen calls "democracy renovation."[2] Renovation does not have to be specific to AI to be responsive to it. Governments must take overdue action to respond to the pressure intensified by AI before it overwhelms our institutions. Just as many unions have organized their members to respond to the risks to labor and exploitative employment relationships exacerbated by AI, so too should political movements galvanize their constituents to respond to the long-standing democratic threats magnified by AI.

The 2023 Writers Guild of America strike, which we discussed previously, provides an excellent model for responding to the threats posed by AI.[3] While members of the Guild still face a precarious future for their profession, the strike was an exemplary demonstration of how to build power using AI, both as a symbol of the professional risks they face and as a practical tool to enhance their success in a changing economy. The writers mobilized to **reform** their conditions in anticipation of the risks posed by AI, secured access for their members to use AI in **responsible** ways, activated widespread public support to **resist** negative uses of AI, and **renovated** their institutions—that is, they strengthened their union—to thrive during challenging circumstances. This is a blueprint for grappling with AI throughout society.

Increasingly advanced AI systems will be created, and people will seek to apply them everywhere they can offer an advantage. It's critical to create conditions for those applications to benefit the public, rather than to concentrate power. Yet the richest elites now benefit from the development and selling of AI. The confluence of AI innovation, a handful of Big Tech firms, venture capital, and lobbying and political power create a positive feedback cycle around AI whereby money begets power begets more money.

We need structural mechanisms to disrupt, or at least to limit, that feedback cycle. One mechanism we have already discussed is the creation of alternatives to corporate AI, distributing control of the technology. Another is a tech impact tax, which returns some of the profits garnered via AI tools for use in reinforcing the democratic systems that those products undermine. A meaningful international tax on AI developers' revenues could disrupt the profit-power feedback loop, fund solutions to global issues like AI-related job loss, help compensate creators of content used in AI training, and support local journalism, government watchdogs, and civic infrastructure. Even a small tax by individual nations could make a significant impact.

International cooperation will be critical to the path forward. Buttressing democracy is not a zero-sum competition. One democratic nation does not gain when another democracy falters, whether it be an ally or a competitor. Science and technology is a global enterprise today, and traditional tools of national economic competition like tariffs and export controls will do little to stem the distribution and impact of digital technologies like AI. Hoarding the latest and greatest computer chips or being the first to unlock next-generation AI capabilities offers only a fleeting advantage, since the technology is constantly improving and trivial to copy and distribute. Rather, decades of international academic cooperation on science, AI, and machine learning has demonstrated that there is much to be gained—even by superpowers—through sharing insights, methods, and best practices across borders.

When developing laws and regulatory frameworks for AI, we urge governments to recognize that the impacts of this technology are contextual. Governments should regulate markets and the behaviors of individuals and organizations broadly and rigorously, and not narrowly target rules to AI in isolation. Governance of AI is not just a matter of technology policy, and should not be regarded as strictly regulating what AI is created and how. As technology changes, rules that apply to AI models of only a certain size or type can become outdated after just a few months.

For example, US President Joe Biden's October 2023 Executive Order on AI applied special restrictions and disclosure requirements to models trained using more than a certain number of computing operations. At the time the order was signed, this very high threshold seemed to directly target the next generation of more powerful and, presumably, more compute-intensive models.[4] However, about a year later, DeepSeek-R1 demonstrated that cutting-edge capabilities could be achieved with computing power one-thirtieth of that level,[5] rendering the original threshold specified in the original regulation meaningless.

Instead of guessing about how AI will evolve, regulations should address specific activities, whether they are performed by humans, AI, or—most likely—a combination thereof. For example, the EU AI Act prohibits the use of AI for social scoring (with many exceptions). However, it exempts existing credit scoring systems—both human and algorithmic—that already pose a risk of bias and discrimination.[6] Rather than establish new rules for social scoring aimed exclusively at AI systems, it would have been better to revisit and strengthen rules for activities like credit scoring that consider how AI might exacerbate preexisting risks.

Finally, AI matters to democracy because it can exacerbate the flaws in systems of governance. More than AI-specific regulation, we need good general governance to resist the concentrations of power, injustices, and authoritarian abuses that AI may intensify.

This is crucial. Very few of the risks we've discussed in this book are unique to AI. Mostly, they are the same risks we humans have faced for centuries; but now they are sped up, scaled up, expanded in scope, and highly sophisticated. They're similar to past digital technologies like the internet, as well as social technologies like organized crime or corporations.

The solutions to AI-fueled democratic risks in domains like campaigning, lobbying, and law enforcement can appear similar to traditional best practices in election law and administration, lobbying disclosure, and criminal justice, because the risks are similar to those we have seen before. This is good news, really. We don't need to bank on new inventions to save our democracy from disruption by AI, but we do need to address existing flaws.[7] By defending our democracy from the risks of AI, we are defending our democracy from other threats. Today, this feels more important than ever.

43

CONCLUSION

Democracy has always been inextricably linked with technology. Whether we like it or not, AI is destined to bring changes to our political life and systems of governance.

To recognize the potential for AI to transform our democracy, you don't have to be a true believer in the potential power of AI tools, or even particularly impressed by them. It's enough to realize that the political class is eagerly adopting these tools in service of their campaigns and governance. AI is already being used to deliver personalized messaging to voters and donors, provide expert assistance to judges, and even write laws.

This is just the first sprint of a trend that will continue throughout the world, and accelerate for years to come. Most of the enduring transformations won't come from the top down but the bottom up. Affluent, influential individuals and those in authority will use AI tools as soon as they are useful; they will not wait to be directed to do so. Politicians, campaign managers, lawyers, and political advocacy groups won't wait for permission. Judges will use AI to draft their decisions because it saves time. News organizations will use AI because it saves money. Tax agencies will add AI to their already algorithmic systems for auditing.

This sort of change has happened many times before. In the nineteenth century, railway networks enabled relatively speedy travel and distribution of goods, which changed the ways that politicians campaigned while transforming the material conditions and social concerns of their constituents. Rail's economic impacts spawned labor movements in the US, UK, and across the world, arguably leading to the Russian Revolution and setting the conditions for monumental shifts in political affairs that resulted in the US Civil War. There is no simple through line that captures all the causes and effects of this transformation; it's not as though rail technology was designed to transform democracy or overthrow governments. Railroads were constructed to make money for their owners and for industrialists of the day. They did, but did not neatly or exclusively serve the interests of any particular person, party, or movement.

We hope the risks associated with AI might inspire people to improve democracy, but that won't happen without deliberate action. If we want to restrain authoritarians, runaway lobbyists, corrupt politicians, and abusive police from using AI to advance their worst impulses, we should urgently reform these institutions and restrain the deleterious behaviors that AI might supercharge. We should strengthen bulwarks against autocracy, advance campaign finance and electioneering regulations, rethink the role of policing in society, and reform endless other aspects of governance.

These are not new problems, and their causes and solutions are not limited to AI. That's our point. In many democracies, good governance reform is long overdue, but has broad-based constituent support, wanting only the political willpower to disrupt the status quo.

Democracy will change, as it always has, to accommodate new technologies. Democracies around the world are currently roiled in debate over what to do about surging inequality, eroding democratic systems, rising authoritarianism, an unstable world order, worsening climate change, and social dislocation. Each polity's response to those threats is not predestined, but it is inevitable that the development and use of AI will influence both the manifestation of and response to each risk.

The essential promise of democracy is that, while no single entity pulls the strings of public life, we all get to give a tug. Collectively, humans will help shape society's next phase of transformation—the first to be influenced by AI. None of us will get exactly the future we want, but we can participate, compromise, and find common ground with each other while advocating for our own rights and needs. Many modern democracies seem to be failing to afford people the agency they deserve in this deliberation, which gives the public even more reason to become engaged and work for change.

We are optimistic about the potential democratic impacts of AI, even though we see its flaws and potential for misuse, because we want to live in a world improved by the cleverness and industry of humankind. We see a landscape of choices ahead and maintain hope that we can navigate that landscape safely together. AI can make democracy more democratic, or more inequitable. Automation can benefit everyone in our society, or just make the rich richer. We are optimistic because it seems possible to reach consensus about the right path.

AI is already magnifying the greatest risks to democracy in the modern world. Society's response should not be limited to focusing on science fiction scenarios of AI Armageddon, or treating AI like the next battlefield in a global cold war; it should entail an urgent reassessment of the frailties and failings of our democratic systems. Many of these challenges have long predated AI, yet the democratic institutions they afflict might not survive the enhanced speed, scale, scope, and sophistication that advancing technology is lending to those undermining them. If we take this opportunity to address the decay in the foundational systems of our democracies, and find clever and responsible uses for AI to shore things up, we may just be able to use this technology to rewire democracy to better serve all of us.

ACKNOWLEDGMENTS

We are merely medium-sized language models made of meat and seawater. As such, we could never have written this book alone.

We would like to thank the MIT Press for publishing the book—specifically, our editor Gita Devi Manaktala, as well as Debora Kuan, David Weinberger, and everyone else who turned our writing into the finished product now in your hands.

Many people helped with this book during the year it took to write. Bruce's research assistants at the Harvard Kennedy School were invaluable: Dom Balasuriya, Eric Gong, Ruthie Gottesman, Dillon Leet, Zubair Merchant, and Lucas Schmuck. Thank you to the participants of the first three International Workshops on Reimagining Democracy. Thank you to all the colleagues and friends who read and commented on the book in various stages of draft: Daron Acemoglu, Slavina Ancheva, Susan Benesch, Kelly Born, Jon Callas, Federica Carugati, Vint Cerf, Wes Chow, Tantum Collins, Jack Conlin, Nick Couldry, Ruthanna Emrys, Melina Geser-Stark, Jordan Hall, Marci Harris, Meghan Harrison, Rose Hendricks, Bill Herdle, Patrick Herron, Stephen

Hilgartner, Jonathan Hochman, Sarah Hubbard, Stephen Kamman, Alexander Klimburg, Gretchen Krueger, Crystal Lee, Hillary Lehr, Ann Lewis, Gideon Lichfield, Matt Marolda, George McArdle, Tom McGee, Adria Meira, Doug Palmer, Eli Pariser, Hugh Pearson, Jennifer Pfister, Barath Raghavan, Nitin Ranjan, Manon Revel, Nathan Schneider, Wendy Seltzer, Ben Shneiderman, Adam Shostack, Vandinika Shukla, Taren Stinebrickner-Kauffman, Sarah Lai Stirland, Mustafa Suleyman, Sheethal Surendran, Danna Thein, Fabian Ulmer, Matt Victor, David Weinberger, Josephine Wolff, and Maria Xynou. A special thank you to our development editor Evelyn Duffy, Kathleen Seidel for her extensive and insightful comments, our copyeditor Beth Friedman, and Cameron Reed for formatting the references. All their feedback made this book—its prose and its ideas—dramatically better.

While we see many useful applications of AI, we wrote this book ourselves. All the ideas and words are our own or stemming from those we have cited or acknowledged.

We thank Tammy, Shannon, Alyana, Bryelle, and all our families who supported us during constant sessions of writing and rewriting. Bruce thanks the Kennedy School of Government: particularly the Ash Center for Democratic Governance and Innovation, and the Belfer Center for Science and International Affairs. Nathan thanks Alicia Soderberg, Denise Provost, Pat Jehlen, Dan Smith, and the MAPLE team, who gave him inspiration and opportunity to grapple with the challenges of legislative participation and the role technology can play. We both thank the Berkman Klein Center for Internet & Society at Harvard University, where we met and germinated our ideas amongst dozens of brilliant scholars thinking deeply about every aspect of democracy and technology addressed in this book. And finally, we thank everyone who is working in these uncertain and dangerous times to preserve and strengthen democracy, both in the US and around the world.

NOTES

PREFACE

1. Kyle Chayka, "Elon Musk's A.I.-Fuelled War on Human Agency," *New Yorker*, February 12, 2025, https://www.newyorker.com/culture /infinite-scroll/elon-musks-ai-fuelled-war-on-human-agency.

2. "Read: JD Vance's Full Speech on AI and the EU," *Spectator*, February 12, 2025, https://www.spectator.co.uk/article/read-jd-vances-full -speech-on-ai-and-the-eu/.

3. Christopher Lehane, "OpenAI's Proposals for the U.S. AI Action Plan," OpenAI, March 13, 2025, https://openai.com/global-affairs /openai-proposals-for-the-us-ai-action-plan.

CHAPTER 1

1. Sterling Dow, "Aristotle, the *Kleroteria*, and the Courts," *Harvard Studies in Classical Philology* 50, no. 1 (1939), https://doi.org/10.2307 /310590.

2. Erin Crochetière, "Democracy and the Lot: The Lottery of Public Offices in Classical Athens" (Master's thesis, McGill University, 2013), https://central.bac-lac.gc.ca/.item?id=TC-QMM-123163&op=pdf &app=Library&oclc_number=91104301.

3. David Runciman, *The Handover: How We Gave Control of Our Lives to Corporations, States, and AIs* (Livewright, 2023).

4. Leonie Cater, "What Estonia's Digital ID Scheme Can Teach Europe," *Politico*, March 13, 2021, https://www.politico.eu/article /estonia-digital-id-scheme-europe/.

5. "Alignment Assemblies," Ministry of Digital Affairs, accessed March 20, 2025, https://moda.gov.tw/en/major-policies/alignment-assemblies/1453.

6. Nathan E. Sanders and Bruce Schneier, "How ChatGPT Hijacks Democracy," *New York Times*, January 15, 2023, https://www.nytimes.com/2023/01/15/opinion/ai-chatgpt-lobbying-democracy.html.

7. OpenAI's ChatGPT (via Gary Apple), "ChatGPT Wrote (Most of) This Letter," *New York Times*, January 24, 2023, https://www.nytimes.com/2023/01/24/opinion/letters/democracy-chatbot.html.

8. Bruce Schneier, "Third Interdisciplinary Workshop on Reimagining Democracy (IWORD 2024)," *Schneier on Security* (blog), February 15, 2025, https://www.schneier.com/blog/archives/2025/01/third-interdisciplinary-workshop-on-reimagining-democracy-iword-2024.html.

9. Alicia Combaz et al., "Applications of Artificial Intelligence Tools to Enhance Legislative Engagement: Case Studies from Make.Org and MAPLE," arXiv, February 12, 2025, https://arxiv.org/abs/2503.04769.

10. Katie Lannan, "Mass. Residents Will Now Be Alerted When Sewage Is Dumped into Local Waterways," *WBUR.org*, February 16, 2021, https://www.wbur.org/news/2021/02/16/baker-signs-bill-sewage-dump-waterways.

CHAPTER 2

1. "Packed with Loopholes: Why the AI Act Fails to Protect Civic Space and the Rule of Law," European Center for Not-for-Profit Law, April 3, 2024, https://ecnl.org/news/packed-loopholes-why-ai-act-fails-protect-civic-space-and-rule-law.

2. Evgeny Morozov, "The AI We Deserve," *Boston Review*, December 4, 2024, https://www.bostonreview.net/forum/the-ai-we-deserve/; Bruce Schneier and Nathan Sanders, "Trust Issues," *Boston Review*, December 4, 2024, https://www.bostonreview.net/forum/the-ai-we-deserve/.

3. Nik Marda et al., "Public AI: Making AI Work for Everyone, by Everyone," Mozilla, September 2024, https://assets.mofoprod.net/network/documents/Public_AI_Mozilla.pdf.

4. Erin Blakemore, "How Photos Became a Weapon in Stalin's Great Purge," *History*, April 11, 2022, https://www.history.com/news/josef-stalin-great-purge-photo-retouching.

5. "The Case of the Moved Body," The Library of Congress, accessed March 20, 2025, https://www.loc.gov/collections/civil-war-glass-neg atives/articles-and-essays/does-the-camera-ever-lie/the-case-of-the -moved-body/.

6. Bruce Schneier and James Waldo, "AI Can Thrive in Open Soci- eties," *Foreign Policy*, June 13, 2019, https://foreignpolicy.com/2019 /06/13/ai-can-thrive-in-open-societies/.

CHAPTER 3

1. Bruce Schneier, "The Internet of Things Will Be the World's Big- gest Robot," *Forbes*, February 2, 2016, https://www.forbes.com/sites /bruceschneier/2016/02/02/the-internet-of-things-will-be-the-worlds -biggest-robot/.

2. Joseph Nocera, "The Day the Credit Card Was Born," *Washington Post*, November 3, 1994, https://www.washingtonpost.com/archive /lifestyle/magazine/1994/11/04/the-day-the-credit-card-was-born /d42da27b-0437-4a67-b753-bf9b440ad6dc/.

3. Alexander Winton, "Get a Horse! America's Skepticism Toward the First Automobiles," *Saturday Evening Post*, July 26, 2022, https://www .saturdayeveningpost.com/2017/01/get-horse-americas-skepticism -toward-first-automobiles/.

4. Ulugbek Vahobjon Ugli Ismatullaev and Sang-Ho Kim, "Review of the Factors Affecting Acceptance of AI-Infused Systems," *Human Factors* 66, no. 1 (2022): 126–44, https://doi.org/10.1177/001872082 11064707.

CHAPTER 4

1. Marcos V. Conde et al., "Real-Time 4K Super-Resolution of Com- pressed AVIF Images. AIS 2024 Challenge Survey," *CVPR 2024 Open Access*, 2024, https://openaccess.thecvf.com/content/CVPR2024W/AI4 Streaming/html/Conde_Real-Time_4K_Super-Resolution_of_Com pressed_AVIF_Images._AIS_2024_Challenge_CVPRW_2024_paper .html.

2. Ben Abraham, "Nvidia's DLSS Carbon Impacts: The Cost-Benefit of Upscaling, Frame Generation, and Neutral Network Training," Green- ing the Games Industry, January 5, 2024, https://gtg.benabraham

.net/nvidia-dlss-carbon-impacts-whats-the-cost-benefit-of-upscaling
-frame-generation-and-neutral/.

3. Simone Angarano et al., "Generative Adversarial Super-Resolution at the Edge with Knowledge Distillation," *Engineering Applications of Artificial Intelligence* 123, part B (2023), https://doi.org/10.1086/725865.

4. Zhiwen Deng et al., "Super-Resolution Reconstruction of Turbulent Velocity Fields Using a Generative Adversarial Network-Based Artificial Intelligence Framework," *Physics of Fluids* 31, no. 12 (2019), https://doi.org/10.1063/1.5127031.

5. Johnny Ryan, "ICCL Report on the Scale of Real-Time Bidding Data Broadcasts in the U.S. and Europe," Irish Council for Civil Liberties, February 5, 2023, https://www.iccl.ie/news/iccl-report-on-the-scale-of-real-time-bidding-data-broadcasts-in-the-u-s-and-europe/.

6. Feng-hsiung Hsu et al., "Deep Blue System Overview," in *ICS '95: Proceedings of the 9th International Conference on Supercomputing,* ed. Laxmi N. Bhuyan et al. (1995), 240–244, https://dl.acm.org/doi/10.1145/224538.224567.

7. Carlos Outeiral, "Current Structure Predictors Are Not Learning the Physics of Protein Folding," *Bioinformatics* 38, no. 7 (2022): 1881–87, e0263069, https://doi.org/10.1093/bioinformatics/btab881.

CHAPTER 5

1. Henry Farrell and Bruce Schneier, "Rechanneling Beliefs: How Information Flows Hinder or Help American Democracy," Johns Hopkins Stavros Niarchos Foundation SNF Agora Institute, May 24, 2021, https://snfagora.jhu.edu/publication/rechanneling-beliefs/.

2. Bruce Schneier, "Rethinking Democracy for the Age of AI," *CyberScoop*, May 28, 2024, https://cyberscoop.com/rethinking-democracy-ai/.

3. Angela Huyue Zhang, "Authoritarian Countries' AI Advantage," *Project Syndicate*, December 23, 2024, https://www.project-syndicate.org/commentary/how-china-uae-became-artificial-intelligence-power houses-by-angela-huyue-zhang-2024-09.

4. Brent Kallmer, "The Road to Digital Unfreedom: How Artificial Intelligence Is Reshaping Repression," *Journal of Democracy* 30, no. 1 (2019): 40–52, https://www.journalofdemocracy.org/articles/the

-road-to-digital-unfreedom-how-artificial-intelligence-is-reshaping -repression/.

5. Helen Toner, talk at the International Workshop on Reimagining Democracy 2024, Washington DC, 11–12 Dec 2024.

CHAPTER 6

1. Baroness Kidron, Debate on Amendment 44A to the Data Use and Access Bill, UK House of Lords, January 28, 2025, https://www.they workforyou.com/lords/?id=2025-01-28c.150.0#g150.2.

CHAPTER 7

1. Joy Buolamwini and Timnit Gebru, "Gender Shades: Intersectional Accuracy Disparities in Commercial Gender Classification," *PMLR* 81 (2018): 1–18, https://proceedings.mlr.press/v81/buolamwini18a.html.

2. Rachael Tatman, "Gender and Dialect Bias in YouTube's Automatic Captions," in *Proceedings of the First ACL Workshop on Ethics in Natural Language Processing*, ed. Dick Hovy et al. (Association for Computational Linguistics, 2017), 53–59, http://www.ethicsinnlp.org /workshop/pdf/EthNLP06.pdf.

3. Jan Betley et al., "Emergent Misalignment: Narrow Finetuning Can Produce Broadly Misaligned LLMs," arXiv, 2025, https://doi.org /10.48550/arXiv.2502.17424.

4. Alina Holmstrom, "Who Is Emma? Using Chatbots in Immigration Services," Application, Ethics, and Governance of AI, March 8, 2023, https://aegai.nd.edu/latest/who-is-emma-using-chatbots-in-immigra tion-services/.

5. Tobias Schimanski et al., "Towards Faithful and Robust LLM Specialists for Evidence-Based Question-Answering," arXiv, 2024, https:// doi.org/10.48550/arXiv.2402.08277.

6. Kyungha Kim et al., "Can LLMs Produce Faithful Explanations for Fact-Checking? Towards Faithful Explainable Fact-Checking via Multi-Agent Debate," arXiv, 2024, https://doi.org/10.48550/arXiv.2402 .07401.

7. "Directive on Automated Decision-Making," Government of Canada, last modified April 25, 2023, https://www.tbs-sct.canada.ca/pol/doc -eng.aspx?id=32592.

CHAPTER 8

1. "Live Speech to Text and Machine Translation Tool for 24 Languages," EU Funding and Tenders Portal, August 5, 2019, https://etendering.ted.europa.eu/cft/cft-documents.html?cftId=5249.

2. Nikki Davidson, "Can AI-Powered Virtual Assistants Crack the Language Barrier?," *Government Technology*, January 23, 2024, https://www.govtech.com/biz/data/can-ai-powered-virtual-assistants-crack-the-language-barrier.

3. Anne Field, "Ukraine Startup Translates Videos for Zelensky, While Adjusting to Work in a War Zone," *Forbes*, July 29, 2022, https://www.forbes.com/sites/annefield/2022/07/29/ukraine-startup-translates-videos-for-zelensky-while-adjusting-to-work-in-a-war-zone/.

4. Jack Nicas and Lucia Cholakian Herrera, "Is Argentina the First A.I. Election?," *New York Times*, November 15, 2023, https://www.nytimes.com/2023/11/15/world/americas/argentina-election-ai-milei-massa.html.

5. Amrita Khalid, "Pakistan's Former Prime Minister Is Using an AI Voice Clone to Campaign from Prison," *Verge*, December 18, 2023, https://www.theverge.com/2023/12/18/24006968/imran-khan-ai-pakistan-prime-minister-voice-clone-elevenlabs.

6. Chance Townsend, "Surprise Democratic Primary Winner Credits AI for Beating Biden," *Mashable*, March 9, 2024, https://mashable.com/article/candidate-uses-ai-credits-it-for-american-samoa-win.

7. "2024 Tokyo Governor Election, Japan," NamuWiki, last modified February 9, 2025, https://en.namu.wiki/w/2024%EB%85%84%20%EC%9D%BC%EB%B3%B8%20%EB%8F%84%EC%BF%84%EB%8F%84%EC%A7%80%EC%82%AC%20%EC%84%A0%EA%B1%B0.

8. Gideon Lichfield, "Meet Your AI Politician of the Future," *Futurepolis*, October 4, 2024, https://futurepolis.substack.com/p/meet-your-ai-politician-of-the-future.

9. Anna Tong and Helen Coster, "Meet Ashley, the World's First AI-Powered Political Campaign Caller," Reuters, December 15, 2023, https://www.reuters.com/technology/meet-ashley-worlds-first-ai-powered-political-campaign-caller-2023-12-12/.

10. Adrian Carrasquillo, "Democrats Use AI in Effort to Stay Ahead with Latino and Black Voters," *Guardian*, August 21, 2024, https://

www.theguardian.com/us-news/article/2024/aug/21/democrats-ai
-black-latino-voters.

11. Ted Brader and Joshua A. Tucker, "Unreflective Partisans? Policy
Information and Evaluation in the Development of Partisanship,"
Political Psychology 39, no. S1 (February 1, 2018): 137–57, https://doi
.org/10.1111/pops.12480.

12. Maurice Jakesch et al., "Co-writing with Opinionated Language
Models Affects Users' Views," arXiv, 2023, https://doi.org/10.1145
/3544548.3581196.

13. Thomas H. Costello et al., "Durably Reducing Conspiracy Beliefs
Through Dialogues with AI," *Science* 385, no. 6714 (September 13,
2024), https://doi.org/10.1126/science.adq1814.

14. Thomas H. Costello et al., "Just the Facts: How Dialogues with
AI Reduce Conspiracy Beliefs," PsyArXiv *Preprints*, February 16, 2025,
https://doi.org/10.31234/osf.io/h7n8u_v1.

15. David Broockman and Joshua Kalla, "Durably Reducing Trans-
phobia: A Field Experiment on Door-to-Door Canvassing," *Science* 352,
no. 6282 (2016): 220–24, https://doi.org/10.1126/science.aad9713;
Joshua L. Kalla and David E. Broockman, "Reducing Exclusionary
Attitudes Through Interpersonal Conversation: Evidence from Three
Field Experiments," *American Political Science Review* 114, no. 2 (2020):
410–25, https://doi.org/10.1017/s0003055419000923.

16. Jackie Mansky, "The Age-Old Problem of 'Fake News,'" *Smithson-
ian Magazine*, May 7, 2018, https://www.smithsonianmag.com/history
/age-old-problem-fake-news-180968945/.

17. Masha Gessen, "The Photo Book That Captured How the Soviet
Regime Made the Truth Disappear," *New Yorker*, July 15, 2018, https://
www.newyorker.com/culture/photo-booth/the-photo-book-that
-captured-how-the-soviet-regime-made-the-truth-disappear.

18. Anat Ben-David and Elinor Carmi, "Dark Cycles: Social Engineering and Political Chatbots in Netanyahu's 2019 Election Campaigns,"
International Journal of Communication 19 (2025): 592–616, https://
ijoc.org/index.php/ijoc/article/view/22420/4906.

19. Editorial Board, "From Iran and Russia, the Disinformation Is
Now. The Target: America," *Washington Post*, August 22, 2024, https://
www.washingtonpost.com/opinions/2024/08/19/russia-iran-ai-dis
information-election/.

20. Domenico Montanaro, "The Truth in Political Advertising: 'You're Allowed to Lie,'" NPR, March 17, 2022, https://www.npr.org /2022/03/17/1087047638/the-truth-in-political-advertising-youre -allowed-to-lie.

CHAPTER 9

1. Andy Brownback and Aaron Novotny, "Social Desirability Bias and Polling Errors in the 2016 Presidential Election," *Journal of Behavioral and Experimental Economics* 74 (2018): 38–56, https://doi.org/10 .1016/j.socec.2018.03.001.

2. Michael A. Bailey, "A New Paradigm for Polling," *Harvard Data Science Review*, July 27, 2023, https://hdsr.mitpress.mit.edu/pub/ejk5yhgv /release/4.

3. Yuxiang Gao et al., "Improving Multilevel Regression and Post-stratification with Structured Priors," *Bayesian Analysis* 16, no. 3 (2020), https://doi.org/10.1214/20-ba1223.

4. Nate Silver, "Calculating 'House Effects' of Polling Firms," *FiveThirtyEight*, June 22, 2012, https://fivethirtyeight.com/features/calculating -house-effects-of-polling-firms/.

5. Kellyton Dos Santos Brito et al., "A Systematic Review of Predicting Elections Based on Social Media Data: Research Challenges and Future Directions," *IEEE Transactions on Computational Social Systems* 8, no. 4 (2021): 819–43, https://doi.org/10.1109/tcss.2021.3063660.

6. Shannon Vallor, *The AI Mirror: How to Reclaim Our Humanity in an Age of Machine Thinking* (Oxford University Press, 2024).

7. "About," Expected Parrot, accessed March 22, 2025, https://www .expectedparrot.com/about.

8. James Bisbee et al., "Synthetic Replacements for Human Survey Data? The Perils of Large Language Models," SocArXiv Papers, August 9, 2023, https://doi.org/10.31235/osf.io/5ecfa; Lisa P. Argyle et al., "Out of One, Many: Using Language Models to Simulate Human Samples," arXiv (Cornell University), September 14, 2022, https://doi .org/10.48550/arxiv.2209.06899.

9. Nathan E. Sanders, Alex Ulinich, and Bruce Schneier, "Demonstrations of the Potential of AI-Based Political Issue Polling," *Harvard Data Science Review* 5, no. 4 (2023), https://doi.org/10.1162/99608f92.1d3 cf75d.

10. Joshua C. Yang et al., "LLM Voting: Human Choices and AI Collective Decision Making," arXiv, August 14, 2024, https://doi.org /10.48550/arxiv.2402.01766; Maud Reveilhac and Davide Morselli, "Article: ChatGPT as a Voting Application in Direct Democracy," *OSF Preprints*, April 18, 2024, https://doi.org/10.31219/osf.io/65v2y; Jairo Gudiño-Rosero et al., "Large Language Models (LLMs) as Agents for Augmented Democracy," arXiv, July 30, 2024, https://arxiv.org/abs /2405.03452.

11. Mengxin Wang et al., "Large Language Models for Market Research: A Data-Augmentation Approach," arXiv, January 6, 2025, https://doi.org/10.48550/arxiv.2412.19363.

CHAPTER 10

1. Alan S. Gerber and Donald P. Green, "Does Canvassing Increase Voter Turnout? A Field Experiment," *Proceedings of the National Academy of Sciences* 96, no. 19 (1999): 10939–42, https://doi.org/10.1073 /pnas.96.19.10939.

2. Robert M. Bond et al., "A 61-Million-Person Experiment in Social Influence and Political Mobilization," *Nature* 489, no. 7415 (2012): 295–98, https://doi.org/10.1038/nature11421.

3. Micah L. Sifry, "How AI Is Transforming the Way Political Campaigns Work," *Nation*, January 31, 2024, https://www.thenation.com /article/politics/how-ai-is-transforming-the-way-political-campaigns -work/.

4. Katie Rogers, "Door Knocking Is Tough These Days. Harris's Team Is Betting on Apps," *New York Times*, October 17, 2024, https://www .nytimes.com/2024/10/17/us/politics/harris-trump-outreach-milwaukee -wisconsin.html.

5. Cat Zakrzewski, "ChatGPT Breaks Its Own Rules on Political Messages," *Washington Post*, August 28, 2023, https://www.washington post.com/technology/2023/08/28/ai-2024-election-campaigns-dis information-ads/.

CHAPTER 11

1. Manoj Sharma, "General Elections 2024: Most Expensive Polls Ever, Anywhere in the World,'" *Fortune India*, May 8, 2024, https://

www.fortuneindia.com/macro/general-elections-2024-most-expen
sive-polls-ever-anywhere-in-the-world/116748.

2. Henry Dyer, "Political Spending and Donations: What Are the Rules in the UK?," *Guardian*, June 6, 2024, https://www.theguardian .com/politics/article/2024/jun/06/political-spending-and-donations -what-are-the-rules-in-the-uk.

3. Joo-Cheong Tham et al., *Digital Campaigning and Political Finance in the Asia and the Pacific Region: A New Age for an Old Problem*, October 14, 2022, https://doi.org/10.31752/idea.2022.37.

4. David Atkin and Elsa Pilichowski, "Regulating Corporate Political Engagement," *Public Governance Policy Papers*, March 18, 2022, https://doi.org/10.1787/8c5615fe-en.

5. Kyle Wiggers, "A Nonprofit Is Using AI Agents to Raise Money for Charity," *TechCrunch*, April 8, 2025. https://techcrunch.com/2025/04 /08/a-nonprofit-is-using-ai-agents-to-raise-money-for-charity/.

6. Tiffany Hsu and Steven Lee Myers, "A.I.'s Use in Elections Sets Off a Scramble for Guardrails," *New York Times*, June 25, 2023, https:// www.nytimes.com/2023/06/25/technology/ai-elections-disinforma tion-guardrails.html.

7. Thiago S. Guzella and Walmir M. Caminhas, "A Review of Machine Learning Approaches to Spam Filtering," *Expert Systems with Applications* 36, no. 7 (2009): 10206–22, https://doi.org/10.1016/j.eswa.2009 .02.037.

8. Adrianne Jeffries et al., "Is Gmail Hiding Bernie's Emails to You? How Inbox Filtering May Impact Democracy," *Guardian*, February 26, 2020, https://www.theguardian.com/us-news/2020/feb/26/gmail-hiding -bernie-sanders-emails-google-inbox-sorting-consequences-2020.

9. Mike Scarcella, "Google Defeats RNC Lawsuit Claiming Email Spam Filters Harmed Republican Fundraising," Reuters, July 31, 2024, https://www.reuters.com/legal/transactional/google-defeats-rnc -lawsuit-claiming-email-spam-filters-harmed-republican-2024-07 -31/.

10. David A. Fahrenthold, Tiff Fehr, and Charlie Smart, "How to Raise $89 Million in Small Donations, and Make It Disappear," *New York Times*, May 14, 2023, https://www.nytimes.com/interactive/2023/05 /14/us/politics/scam-robocalls-donations-policing-veterans.html.

11. California Business and Professions Code, BPC 19741, https://leginfo.legislature.ca.gov/faces/codes_displaySection.xhtml?sectionNum=17941.&lawCode=BPC.

12. "Regulation (EU) 2024/1689 of the European Parliament and of the Council of 13 June 2024," European Union, June 13, 2024, https://eur-lex.europa.eu/eli/reg/2024/1689/oj/eng.

13. Alix Martichoux, "Why Being on the 'Do Not Call' List Doesn't Actually Stop Spam Calls, Texts," *The Hill*, March 10, 2024, https://thehill.com/homenews/nexstar_media_wire/4507529-why-being-on-the-do-not-call-list-doesnt-actually-stop-spam-calls-texts/.

14. Mike Isaac, "Facebook Ends Ban on Political Advertising," *New York Times*, March 3, 2021, https://www.nytimes.com/2021/03/03/technology/facebook-ends-ban-on-political-advertising.html.

15. Julia Angwin and Jeff Larson, "Help Us Monitor Political Ads Online," *ProPublica*, September 7, 2017, https://www.propublica.org/article/help-us-monitor-political-ads-online.

CHAPTER 12

1. Dan Rosenzweig-Ziff and Jenna Sampson, "Mayoral Candidate Vows to Let VIC, an AI Bot, Run Wyoming's Capital City," *Washington Post*, August 19, 2024, https://www.washingtonpost.com/technology/2024/08/19/artificial-intelligence-mayor-cheyenne-vic/.

2. "Laramie County, Wyoming Primary Election Summary Results Report," Laramie County, WY Government, August 20, 2024, https://www.laramiecountywy.gov/files/sharedassets/public/v/4/clerk/documents/election-results/2024primaryresults.pdf.

3. Angela Yang and Daniele Hamamdjian, "AI Candidate Running for Parliament in the UK Says AI Can Humanize Politics," NBC News, June 13, 2024, https://www.nbcnews.com/tech/tech-news/ai-candidate-running-parliament-uk-says-ai-can-humanize-politics-rcna156991.

4. TOI World Desk, "UK General Election: AI Candidate Gets Only 179 Votes, Finishes Last," *Times of India*, July 5, 2024, https://timesofindia.indiatimes.com/world/uk/uk-general-election-ai-candidate-gets-only-179-votes-finishes-last/articleshow/111508515.cms.

5. Greg Bensinger, "Virginia Congressional Candidates Debate Incumbent's AI—with a Few Glitches," Reuters, October 18, 2024, https://

www.reuters.com/technology/artificial-intelligence/virginia-congres
sional-candidates-debate-incumbents-ai-with-few-glitches-2024-10-18/.

6. Bruce Schneier and Nathan Sanders, "Six Ways That AI Could
Change Politics," *MIT Technology Review*, July 26, 2023, https://www
.technologyreview.com/2023/07/28/1076756/six-ways-that-ai-could
-change-politics/.

CHAPTER 13

1. Brandon L. Garrett and Cynthia Rudin, "The Right to a Glass Box:
Rethinking the Use of Artificial Intelligence in Criminal Justice,"
Duke Law School Public Law & Legal Theory Series No. 2023-03, Feb-
ruary 17, 2023, https://doi.org/10.2139/ssrn.4275661.

2. Emily Pronin, "The Introspection Illusion," in *Advances in Experimental Social Psychology Volume 41*, ed. Mark P. Zanna (Academic
Press, 2009), 1–67, https://doi.org/10.1016/S0065-2601(08)00401-2.

3. Gillian K. Hadfield, "Explanation and Justification: AI Decision-
Making, Law, and the Rights of Citizens," Schwartz Reisman Institute,
May 18, 2021, https://srinstitute.utoronto.ca/news/hadfield-justifiable
-ai.

4. Joyce Zhou and Thorsten Joachims, "How to Explain and Justify
Almost Any Decision: Potential Pitfalls for Accountability in AI
Decision-Making," *2022 ACM Conference on Fairness, Accountability, and
Transparency*, June 12, 2023, 12–21, https://doi.org/10.1145/3593013
.3593972.

CHAPTER 14

1. Maria Yagoda, "Airline Held Liable for Its Chatbot Giving Passenger
Bad Advice—What This Means for Travellers," BBC, February 23, 2024,
https://www.bbc.com/travel/article/20240222-air-canada-chatbot
-misinformation-what-travellers-should-know.

2. Alvaro M. Bedoya, "Statement of Commissioner Alvaro M. Bedoya
on FTC v. Rite Aid Corporation," Federal Trade Commission, December
19, 2023, https://www.ftc.gov/legal-library/browse/cases-proceedings/
public-statements/statement-commissioner-alvaro-m-bedoya-ftc-v
-rite-aid-corporation.

3. Hal Ashton, "Definitions of Intent Suitable for Algorithms," *Artificial Intelligence and Law* 31, no. 3 (2023): 515–46, https://doi.org/10.1007/s10506-022-09322-x.

4. Office of Public Affairs, "Volkswagen AG Agrees to Plead Guilty and Pay $4.3 Billion in Criminal and Civil Penalties," press release, January 11, 2027, https://www.justice.gov/archives/opa/pr/volkswagen-ag-agrees-plead-guilty-and-pay-43-billion-criminal-and-civil-penalties-six.

5. Danielle Kaye et al., "U.S. Accuses Software Maker RealPage of Enabling Collusion on Rents," *New York Times*, August 23, 2024, https://www.nytimes.com/2024/08/23/business/economy/realpage-doj-antitrust-suit-rent.html.

CHAPTER 15

1. Jim Macnamara, *Organizational Listening: The Missing Essential in Public Communication* (Peter Lang, 2016).

2. Beth Simone Noveck, "How AI Could Restore Our Faith in Democracy," *Fast Company*, January 9, 2024, https://www.fastcompany.com/91001497/ai-faith-in-democracy.

3. "AI & Democracy: Insights from Make.org Roundtable on AI Tools," Make.org Blog, December 17, 2024, https://about.make.org/articles-be/ai-democracy-insights-from-make-org-roundtable-on-ai-tools.

4. "MAPLE: Massachusetts Platform for Legislative Engagement," accessed March 22, 2025, https://www.mapletestimony.org/.

5. Committee on House Administration Subcommittee on Modernization, Flash Report: Artificial Intelligence Strategy & Implementation, December 18, 2023, https://cha.house.gov/_cache/files/1/9/19743fef-f6b0-4b12-820c-0f90d4752e9d/86C261E34CEFA60C958EF79D6D852BCB.cha-modernization-ai-flash-report-12-18-23-7-.pdf.

6. "Used AI to Prepare India's Plan for the Next 25 Years, Says PM Modi," *Economic Times*, April 15, 2024, https://economictimes.indiatimes.com/news/elections/lok-sabha/india/used-ai-to-prepare-indias-plan-for-the-next-25-years-says-pm-modi/articleshow/109317177.cms.

7. Sarah Kreps and Maurice Jakesch, "Can AI Communication Tools Increase Legislative Responsiveness and Trust in Democratic Institu-

tions?," *Government Information Quarterly* 40, no. 3 (2023), https://doi
.org/10.1016/j.giq.2023.101829.

8. "Fake Comments: How U.S. Companies & Partisans Hack Democracy to Undermine Your Voice," New York State Office of the Attorney General Letitia James, https://ag.ny.gov/sites/default/files/reports
/oag-fakecommentsreport.pdf.

CHAPTER 16

1. Diane Jeantet and Mauricio Savarese, "Brazilian City Enacts an Ordinance That Was Written by ChatGPT," AP News, November 30, 2023, https://apnews.com/article/brazil-artificial-intelligence-porto
-alegre-5afd1240afe7b6ac202bb0bbc45e08d4.

2. Richard L. Hall and Alan V. Deardorff, "Lobbying as Legislative Subsidy," *American Political Science Review* 100, no. 1 (2006): 69–84, https://doi.org/10.1017/s0003055406062010.

3. Joshua M. Jansa et al., "Copy and Paste Lawmaking: Legislative Professionalism and Policy Reinvention in the States," *American Politics Research* 47, no. 4 (2019): 739–67, https://doi.org/10.1177/1532673
x18776628.

4. "About Xcential Legislative Technologies," Xcential, accessed March 22, 2025, https://xcential.com/about/.

5. "The Comparative Print Suite," POPVOX Foundation, February 27, 2023, https://www.popvox.org/legitech/comparative-print-suite.

6. Joseph Gesnouin et al., "LLaMandement: Large Language Models for Summarization of French Legislative Proposals," arXiv, January 29, 2024, https://doi.org/10.48550/arxiv.2401.16182.

7. Bússola Tech and Luís Kimaid, "Integrating Artificial Intelligence to Legislative Services—Ulysses Suite in the Chamber of Deputies of Brazil," LegisTech Library, March 4, 2024, https://library.bussola-tech
.co/p/ulysses-chamber-deputies-brazil.

8. Chloe Cornish, "UAE Set to Use AI to Write Laws in World First," *Financial Times*, April 20, 2025. https://www.ft.com/content/9019cd51
-2b55-4175-81a6-eafcf28609c3.

9. Corinna Coupette et al., "Law Smells," *Artificial Intelligence and Law* 31, no. 2 (2023): 335–68, https://doi.org/10.1007/s10506-022-09315-w.

10. Robert Pear, "Four Words That Imperil Health Care Law Were All a Mistake, Writers Now Say," *New York Times*, May 26, 2015, https://www.nytimes.com/2015/05/26/us/politics/contested-words-in-afford able-care-act-may-have-been-left-by-mistake.html.

11. Luís Kimaid, "From Paper to Tokens: Transforming Legislative Services in the Chamber of Deputies of Chile," LegisTech Library, February 10, 2025, https://library.bussola-tech.co/p/caminar-ai-camara -diputadas-diputados-chile.

12. "Forecasting GDP with Explainable AI," Observatory of Public Sector Innovation, November 22, 2023, https://oecd-opsi.org/innova tions/forecasting-gdp/.

13. "Quorum Copilot," Quorum, accessed March 24, 2025, https://www.quorum.us/products/copilot/.

14. Bruce Schneier, "The Coming AI Hackers," The Belfer Center for Science and International Affairs, April 2021, https://www.belfer center.org/publication/coming-ai-hackers.

15. Amy Melissa McKay, "Buying Amendments? Lobbyists' Campaign Contributions and Microlegislation in the Creation of the Affordable Care Act," *Legislative Studies Quarterly* 45, no. 2 (2020): 327–60, https://doi.org/10.1111/lsq.12266.

16. Ittai Bar-Siman-Tov, "An Introduction to the Comparative and Multidisciplinary Study of Omnibus Legislation," in *Comparative Multidisciplinary Perspectives on Omnibus Legislation*, ed. Ittai Bar-Siman-Tov (Springer, 2021), https://www.researchgate.net/publication/3519 11699_An_Introduction_to_the_Comparative_and_Multidisciplinary _Study_of_Omnibus_Legislation.

17. Theo Jans and Sonia Piedrafita, "The Role of National Parliaments in European Decision-Making," EIPAScope 2009/1 (2009), 19–26, https://aei.pitt.edu/12376/1/20090709111616_Art3_Eipascoop2009_01.pdf.

18. Kristen Underhill, "Broken Experimentation, Sham Evidence-Based Policy," *Yale Law and Policy Review* 38, no. 1 (2019): 150, https://scholarship.law.columbia.edu/faculty_scholarship/3276.

CHAPTER 17

1. Richard Ngo, talk at the International Workshop on Reimagining Democracy 2024, Washington DC, December 11, 2024.

2. Margaret Wood, "The Revised Statutes of the United States: Predecessor to the U.S. Code," in *Custodia Legis* (blog), July 2, 2015, https://blogs.loc.gov/law/2015/07/the-revised-statutes-of-the-united-states-predecessor-to-the-u-s-code/.

3. "Statute Law Revision Act 2007," Act of the Oireachtas of Ireland number 28 of 2007, https://www.irishstatutebook.ie/eli/2007/act/28/enacted/en/pdf; "Statute Law Revision Act 2007," Wikipedia, accessed March 24, 2025, https://en.wikipedia.org/wiki/Statute_Law_Revision_Act_2007.

4. Senate Government Oversight Committee O.R.C. 101.64 Licensure Review, (2024) (testimony of Joseph Baker, Common Sense Initiative Director), https://search-prod.lis.state.oh.us/api/v2/general_assembly_135/committees/cmte_s_govt_1/meetings/cmte_s_govt_1_2024-05-22-1030_1242/submissions/olr_testimony_csi.pdf.

5. Jon Husted, "AI Can Be a Force for Deregulation," *Wall Street Journal*, March 23, 2025. https://www.wsj.com/opinion/ai-can-be-a-force-for-deregulation-technology-government-ohio-federal-365ed0d4.

6. Catherine M. Sharkey, "AI for Retrospective Review," *Belmont Law Review* 8, no. 2 (2021): 21–46, https://papers.ssrn.com/sol3/papers.cfm?abstract_id=3927987.

CHAPTER 18

1. Congressional Research Service, "Counting Regulations: An Overview of Rulemaking, Types of Federal Regulations, and Pages in the Federal Register," updated September 3, 2019, https://sgp.fas.org/crs/misc/R43056.pdf.

2. *The Brussels Effect: How the European Union Rules the World* (Oxford University Press, 2019), chap. 2, https://doi.org/10.1093/oso/9780190088583.003.0003.

3. David Epstein and Sharyn O'Halloran, "Divided Government and the Design of Administrative Procedures: A Formal Model and Empirical Test," *Journal of Politics* 58, no. 2 (1996): 373–97, https://doi.org/10.2307/2960231.

4. Jason Webb Yackee and Susan Webb Yackee, "Divided Government and US Federal Rulemaking," *Regulation & Governance* 3, no. 2 (2009): 128–44, https://doi.org/10.1111/j.1748-5991.2009.01051.x.

5. Robert Shaffer, "Power in Text: Implementing Networks and Institutional Complexity in American Law," *Journal of Politics* 84, no. 1 (2022): 86–100, https://doi.org/10.1086/714933.

6. Luca Repetto and Maximiliano Sosa Andrés, "Divided Government, Polarization, and Policy: Regression-Discontinuity Evidence From US States," *European Journal of Political Economy* 80 (2023): Article 102473, https://doi.org/10.1016/j.ejpoleco.2023.102473.

7. Congressional Research Service, "No More Deference: Sixth Circuit Relies on Loper Bright to Strike Down Net Neutrality Rules," February 3, 2025, https://crsreports.congress.gov/product/pdf/LSB/LSB11264.

8. Carol S Weissert et al., "Governors in Control: Executive Orders, State-Local Preemption, and the COVID-19 Pandemic," *Publius: The Journal of Federalism* 51, no. 3 (2021): 396–428, https://doi.org/10.1093/publius/pjab013.

9. Thomas Poguntke and Paul Webb, "The Presidentialization of Politics," 2005, https://doi.org/10.1093/0199252017.001.0001.

10. OECD, "OECD Survey on Drivers of Trust in Public Institutions—2024 Results," OECD Publishing, 2024, https://doi.org/10.1787/9a20554b-en.

CHAPTER 19

1. Thomas Poguntke and Paul Webb, eds., *The Presidentialization of Politics: A Comparative Study of Modern Democracies* (Oxford University Press, 2005), https://doi.org/10.1093/0199252017.001.0001; Lan Zhang et al., "A Machine Learning-Based Defensive Alerting System Against Reckless Driving in Vehicular Networks," *IEEE Transactions on Vehicular Technology* 68, no. 12 (2019): 12227–38, https://doi.org/10.1109/tvt.2019.2945398.

2. Michael Torrice, "Ketamine Is Revolutionizing Antidepressant Research, but We Still Don't Know How It Works," *Chemical & Engineering News*, January 15, 2020, https://cen.acs.org/biological-chemistry/neuroscience/Ketamine-revolutionizing-antidepressant-research-still/98/i3.

CHAPTER 20

1. Donna Lu, "We Tried Out DeepSeek. It Worked Well, Until We Asked It About Tiananmen Square and Taiwan," *Guardian*, January

28, 2025, https://www.theguardian.com/technology/2025/jan/28/we-tried-out-deepseek-it-works-well-until-we-asked-it-about-tiananmen-square-and-taiwan.

2. Meredith Broussard, *More than a Glitch: Confronting Race, Gender, and Ability Bias in Tech* (MIT Press, 2024).

3. Jim Henry, "More Drivers Accept Monitoring, to Get Safe-Driver Discounts on Auto Insurance," *Forbes*, May 30, 2022, https://www.forbes.com/sites/jimhenry/2022/05/30/more-drivers-accept-monitoring-to-get-safe-driver-discounts-on-auto-insurance/.

4. Daphne Zhang, "Insurers' AI Use for Coverage Decisions Targeted by Blue States," Bloomberg Law, November 30, 2023, https://news.bloomberglaw.com/insurance/insurers-ai-use-for-coverage-decisions-targeted-by-blue-states.

5. Gerrit De Vynck and Will Oremus, "As AI Booms, Tech Firms Are Laying Off Their Ethicists," *Washington Post*, updated March 30, 2023, https://www.washingtonpost.com/technology/2023/03/30/tech-companies-cut-ai-ethics/.

6. Cory Doctorow, "Our Neophobic, Conservative AI Overlords Want Everything to Stay the Same," *Los Angeles Review of Books*, January 1, 2020, https://lareviewofbooks.org/blog/provocations/neophobic-conservative-ai-overlords-want-everything-stay/.

7. Katherine Bersch and Gabriela Lotta, "Political Control and Bureaucratic Resistance: The Case of Environmental Agencies in Brazil," *Latin American Politics and Society* 66, no. 1 (2024): 27–50, https://doi.org/10.1017/lap.2023.22.

8. Margery Austin Turner et al., "Discrimination in Metropolitan Housing Markets: Phase 1," March 30, 2005, https://www.huduser.gov/portal/publications/hsgfin/hds_phase1.html.

CHAPTER 21

1. Jakob Nielsen, "AI Tools Raise the Productivity of Customer-Support Agents," Nielsen Norman Group, July 16, 2023, https://www.nngroup.com/articles/ai-productivity-customer-support/.

2. Jenny T. Liang, Chenyang Yang, and Brad A. Myers, "A Large-Scale Survey on the Usability of AI Programming Assistants: Successes and Challenges," arXiv, 2023, https://doi.org/10.48550/arxiv.2303.17125.

3. Daniele Checchi et al., "Public Sector Jobs: Working in the Public Sector in Europe and the US," IZA Discussion Papers 14514, https://docs.iza.org/dp14514.pdf.

CHAPTER 22

1. "City of Boston Partners with Google on Traffic Signal Optimization," City of Boston, last updated August 8, 2024, https://www.boston.gov/news/city-boston-partners-google-traffic-signal-optimization.

2. Sharon Jayson, "Why Social Security Disability Claims Are Taking So Long," AARP, January 17, 2024, https://www.aarp.org/retirement/social-security/info-2024/disability-claim-wait-times.html.

3. May Bulman, "More Than 17,000 Sick and Disabled People Have Died While Waiting for Welfare Benefits, Figures Show," *Independent*, July 2, 2019, https://www.independent.co.uk/news/uk/home-news/pip-waiting-time-deaths-disabled-people-die-disability-benefits-personal-independence-payment-dwp-a8727296.html.

4. Virginia Eubanks, *Automating Inequality: How High-Tech Tools Profile, Police, and Punish the Poor* (St. Martin's Press, 2018).

5. Jennifer Pahlka, *Recoding America: Why Government Is Failing in the Digital Age and How We Can Do Better* (Metropolitan Books, 2023).

6. Adam Stone, "Q&A: Microsoft Copilot Should Fly to Agencies This Summer," *FedTech*, March 1, 2024, https://fedtechmagazine.com/article/2024/03/qa-microsoft-copilot-should-fly-to-agencies-summer.

7. Hayden Field, "OpenAI Launches ChatGPT Gov for U.S. Government Agencies," CNBC, January 28, 2025, https://www.cnbc.com/2025/01/28/openai-launches-chatgpt-gov-for-us-government-agencies.html.

8. Frances Mao, "Robodebt: Illegal Australian Welfare Hunt Drove People to Despair," *BBC*, July 7, 2023, https://www.bbc.com/news/world-australia-66130105.

9. Todd Feathers, "Judge Rules $400 Million Algorithmic System Illegally Denied Thousands of People's Medicaid Benefits," *Gizmodo*, August 29, 2024, https://gizmodo.com/judge-rules-400-million-algorithmic-system-illegally-denied-thousands-of-peoples-medicaid-benefits-2000492529.

10. Arvind Narayanan and Sayash Kapoor, *AI Snake Oil: What Artificial Intelligence Can Do, What It Can't, and How to Tell the Difference* (Princeton University Press, 2024).

11. Angelina Wang et al., "Against Predictive Optimization: On the Legitimacy of Decision-Making Algorithms That Optimize Predictive Accuracy," *ACM Journal on Responsible Computing* 1, no. 1 (2024): 1–45, https://doi.org/10.1145/3636509.

12. "Fight Health Insurance," accessed March 25, 2025, https://fight healthinsurance.com/.

13. Kevin De Liban, "Inescapable AI: The Ways AI Decides How Low-Income People Work, Live, Learn, and Survive," TechTonic Justice, November 2024, https://www.techtonicjustice.org/reports /inescapable-ai.

14. OECD, "Tax Administration 2022: Comparative Information on OECD and Other Advanced and Emerging Economies," OECD Publishing, June 23, 2022, https://doi.org/10.1787/1e797131-en.

15. Martyn Landi, "Government Launches Trial of Generative AI Chatbot on Gov.UK," *Independent*, November 5, 2024, https://www .independent.co.uk/business/government-launches-trial-of-generative -ai-chatbot-on-gov-uk-b2641548.html.

16. Canadian Intellectual Property Office, "ISED Business Assistant— Frequently Asked Questions," Government of Canada, last modified September 14, 2020, https://ised-isde.canada.ca/site/canadian-intellec tual-property-office/en/ised-business-assistant-frequently-asked -questions.

17. *Harnessing AI to Improve Government Services and Customer Experience*, 118 Cong. (2024) (written testimony of Beth Simon Noveck, Chief Innovation Officer, The State of New Jersey), https://www .hsgac.senate.gov/wp-content/uploads/Testimony-Noveck-2024-01 -10-REVISED.pdf.

18. Akshaya Suresh and Vinay Viswambharan, "ML And GIS Aid Disaster Response," *ArcUser*, Spring 2022, https://www.esri.com/about /newsroom/arcuser/ml-aids-geospatial-assessment-for-disaster -response/.

19. Nikki Henderson, "How Geospatial Imaging and IT Inform FEMA's Disaster Response," GovCIO Media & Research, January 12,

2023, https://govciomedia.com/how-geospatial-imaging-and-it-inform -femas-disaster-response/.

20. "Artificial Intelligence Use Case Inventory," US Department of Homeland Security, last modified December 16, 2024, https://www .dhs.gov/archive/data/AI_inventory.

21. CAS Science Team, "Addressing Sustainability of the Global Patent System: The Role of AI in Enhancing Productivity," CAS, September 22, 2022, https://www.cas.org/resources/cas-insights/ai-patent -examination-productivity.

22. Kathy Van Der Herten, "Patent Office Sustainability and the Role of Artificial Intelligence," *WIPO Magazine*, January 9, 2023, https:// www.wipo.int/en/web/wipo-magazine/articles/patent-office-sustain ability-and-the-role-of-artificial-intelligence-56004.

23. Justin Doubleday, "FOIA Backlogs on the Rise After Record Number of Requests," *Federal News Network*, March 3, 2023, https:// federalnewsnetwork.com/agency-oversight/2023/03/foia-backlogs -on-the-rise-after-record-number-of-requests/.

24. Scoop News Group, "How the State Department Used AI and Machine Learning to Revolutionize Records Management," *FedScoop*, May 16, 2024, https://fedscoop.com/how-the-state-department-used -ai-and-machine-learning-to-revolutionize-records-management/.

CHAPTER 23

1. Colin Van Noordt and Gianluca Misuraca, "Artificial Intelligence for the Public Sector: Results of Landscaping the Use of AI in Government Across the European Union," *Government Information Quarterly* 39, no. 3 (2022): article 101714, https://doi.org/10.1016/j.giq.2022.101714.

2. Madison Alder, "Federal Government Discloses More Than 1,700 AI Use Cases," *FedScoop*, December 18, 2024, https://fedscoop.com /federal-government-discloses-more-than-1700-ai-use-cases/.

3. "2024 Federal Agency AI Use Case Inventory," Office of the Federal Chief Information Officer, accessed March 25, 2025, https://github .com/ombegov/2024-Federal-AI-Use-Case-Inventory/tree/4a29d1322 91261d4829b187acd8af618e7295294.

4. *A Map of Automated Decision-Making in the NSW Public Sector: A Special Report to Parliament*, New South Wales Ombudsman, March 8,

2024, https://www.ombo.nsw.gov.au/reports/report-to-parliament/a
-map-of-automated-decision-making-in-the-nsw-public-sector-a
-special-report-to-parliament.

5. Clare Martorana, "Why the American People Deserve a Digital
Government," The White House, September 22, 2023, https://web
.archive.org/web/20241220012523/https://www.whitehouse.gov
/omb/briefing-room/2023/09/22/why-the-american-people-deserve-a
-digital-government/.

6. Faiz Surani et al., "AI For Scaling Legal Reform: Mapping and
Redacting Racial Covenants in Santa Clara County," arXiv, February
12, 2025, https://doi.org/10.48550/arXiv.2503.03888; Gideon Lichfield,
"How AI Slashed 40 Years of Bureaucracy to Six Days," *Futurepolis*
(blog), October 18, 2024, https://futurepolis.substack.com/p/how-ai
-slashed-40-years-of-bureaucracy.

7. "Pairing with AI for Public Sector Impact in Singapore," UNDP
Singapore Global Centre, September 26, 2024, https://www.undp
.org/policy-centre/singapore/blog/pairing-ai-public-sector-impact
-singapore.

8. Association of Certified Fraud Examiners, *Occupational Fraud 2022:
A Report to the Nations*, https://legacy.acfe.com/report-to-the-nations
/2022/.

9. Ke Wang et al., "An AI-Based Automated Continuous Compliance
Awareness Framework (CoCAF) for Procurement Auditing," *Big Data
and Cognitive Computing* 4, no. 3 (September 3, 2020), https://doi.org
/10.3390/bdcc4030023.

10. "Treasury Announces Enhanced Fraud Detection Processes,
Including Machine Learning AI, Prevented and Recovered Over $4
Billion in Fiscal Year 2024," U.S. Department of the Treasury, Febru-
ary 8, 2025, https://home.treasury.gov/news/press-releases/jy2650.

11. Richard Fikes and Tom Garvey, "Knowledge Representation and
Reasoning—a History of DARPA Leadership," *AI Magazine* 41, no. 2
(2020): 9–21, https://doi.org/10.1609/aimag.v41i2.5295; Sara Reese
Hedberg, "DART: Revolutionizing Logistics Planning," *IEEE Intelligent
Systems* 17, no. 3 (2002): 81–83, https://doi.org/10.1109/mis.2002.100
5635.

12. "Artificial Intelligence for Logistics Program," National Research
Council Canada, last modified December 24, 2024, https://nrc.canada

.ca/en/research-development/research-collaboration/programs/artificial-intelligence-logistics-program.

13. Joseph Kuba Nembe et al., "The Role of Artificial Intelligence in Enhancing Tax Compliance and Financial Regulation," *Finance & Accounting Research Journal* 6, no. 2 (2024): 241–51, https://doi.org/10.51594/farj.v6i2.822.

14. Janna Brancolini, "Italy Turns to AI to Find Taxes in Cash-First, Evasive Culture," *Bloomberg Tax*, October 31, 2022, https://news.bloombergtax.com/daily-tax-report-international/italy-turns-to-ai-to-find-taxes-in-cash-first-evasive-culture.

15. Beril Akman, "Turkey Turns to AI to Crack Down on Rampant Tax Evasion," *Bloomberg*, July 1, 2024, https://www.bloomberg.com/news/articles/2024-07-01/turkey-turns-to-ai-to-crack-down-on-rampant-tax-evasion?srnd=technology-ai.

16. Oyebola Okunogbe and Gabriel Tourek, "How Can Lower-Income Countries Collect More Taxes? The Role of Technology, Tax Agents, and Politics," *Journal of Economic Perspectives* 38, no. 1 (2024): 90–91, https://doi.org/10.1257/jep.38.1.81.

17. Andres B. Schwarzenberg, "U.S. Government Procurement and International Trade," Congressional Research Service, last updated August 9, 2024, https://www.congress.gov/crs-product/IF11580.

18. Federico Bianchi et al., "How Well Can LLMs Negotiate? NegotiationArena Platform and Analysis," arXiv, February 8, 2024, https://doi.org/10.48550/arxiv.2402.05863.

19. Jinghua Wu et al., "Emotion-Driven Reasoning Model for Agent-Based Human–Computer Negotiation," *Expert Systems with Applications* 240 (2024): article 122448, https://doi.org/10.1016/j.eswa.2023.122448.

20. Matthew Hutson, "AI Learns the Art of Diplomacy," *Science* 378, no. 6622 (2022): 818, https://www.science.org/content/article/ai-learns-art-diplomacy-game.

21. Michela Guida et al., "The Role of Artificial Intelligence in the Procurement Process: State of the Art and Research Agenda," *Journal of Purchasing and Supply Management* 29, no. 2 (2023): article 100823, https://doi.org/10.1016/j.pursup.2023.100823; Sandra Erwin, "Startup Pitches AI Tool to Prevent Pentagon Procurement Blunders," *SpaceNews*, February 14, 2024.

22. "Electronic Quoter: A Purchasing Tool for Efficient Market Research in Peru," OPSI Observatory of Public Sector Innovation, November 16, 2023, https://oecd-opsi.org/innovations/electronic-quoter-peru/.

23. William Press, "At Lunch with Freeman Dyson," *Inference* 6, no. 1 (2021), https://doi.org/10.37282/991819.21.1.

24. Yao Fu et al., "Improving Language Model Negotiation with Self-Play and In-Context Learning from AI Feedback," arXiv, May 17, 2023, https://doi.org/10.48550/arxiv.2305.10142.

25. Samuel "Mooly" Dinnar et al., "Artificial Intelligence and Technology in Teaching Negotiation," *Negotiation Journal* 37, no. 1 (2021): 65–82, https://doi.org/10.1111/nejo.12351.

26. Tinglong Dai, Katia Sycara, and Ronghuo Zheng, "Agent Reasoning in AI-Powered Negotiation," in *Handbook of Group Decision and Negotiation*, ed. D. Marc Kilgour and Colin Eden (Springer, 2021), 1187–1211, https://doi.org/10.1007/978-3-030-49629-6_26.

27. Jinghua Wu et al., "Emotion-Driven Reasoning Model for Agent-Based Human–Computer Negotiation," *Expert Systems with Applications* 240 (2024): article 122448, https://doi.org/10.1016/j.eswa.2023.122448.

28. Federico Bianchi et al., "How Well Can LLMs Negotiate? NegotiationArena Platform and Analysis," arXiv, February 8, 2024, https://doi.org/10.48550/arxiv.2402.05863.

CHAPTER 24

1. Paul Mozur, "One Month, 500,000 Face Scans: How China Is Using A.I. to Profile a Minority," *New York Times*, April 14, 2019, https://www.nytimes.com/2019/04/14/technology/china-surveillance-artificial-intelligence-racial-profiling.html.

2. Kanishka Singh, "Rights Advocates Concerned by Reported US Plan to Use AI to Revoke Student Visas," Reuters, March 6, 2025, https://www.reuters.com/technology/artificial-intelligence/us-use-ai-revoke-visas-students-perceived-hamas-supporters-axios-reports-2025-03-06/.

3. M.H. Lee, "Automated Traffic Cameras Prove Effective in Reducing Accidents in Seoul," *Korea Bizwire*, February 29, 2024, http://koreabizwire.com/automated-traffic-cameras-prove-effective-in-reducing-accidents-in-seoul/274066.

4. "Post Office Horizon Scandal: Why Hundreds Were Wrongly Prosecuted," *BBC News*, July 30, 2024, https://www.bbc.com/news /business-56718036.

5. Kasandra Brabaw, "Why We Don't Want Algorithms to Make Moral Choices," *Chicago Booth Review*, May 11, 2022, https://www .chicagobooth.edu/review/why-we-dont-want-algorithms-make -moral-choices.

6. Rune Elvik, "Risk of Apprehension for Road Traffic Law Violations in Norway," *Accident Analysis & Prevention* 209 (2025): article 107831, https://doi.org/10.1016/j.aap.2024.107831.

7. Sou Hee Yang, "The Implications of Using Digital Technologies in the Management of COVID-19: Comparative Study of Japan and South Korea," *Journal of Medical Internet Research* 25 (2023): article e45705, https://doi.org/10.2196/45705.

8. Stephen Nessen, "MTA Banned from Using Facial Recognition to Enforce Fare Evasion," *Gothamist*, April 28, 2024, https://gothamist .com/news/mta-banned-from-using-facial-recognition-to-enforce -fare-evasion.

9. Richard Diamond, "France Leads World in Speed Camera Opposition," *TheNewspaper.com*, October 23, 2023, https://www.thenews paper.com/news/72/7211.asp.

10. Julie Zauzmer Weil, "IRS: Fund Us, and We'll Collect Hundreds of Billions from Tax Cheaters," *Washington Post*, February 6, 2024, https://www.washingtonpost.com/business/2024/02/06/irs-tax-eva sion-study-budget/.

11. Lance Eliot, "Self-Driving Cars Spurring Nearby Human Drivers into Speeding and Abysmal Reckless Driving," *Forbes*, January 28, 2021, https://www.forbes.com/sites/lanceeliot/2021/01/28/self-driving -cars-spurring-nearby-human-drivers-into-speeding-and-abysmal -reckless-driving/.

CHAPTER 25

1. David F. Drake and Robin L. Just, "Ignore, Avoid, Abandon, and Embrace: What Drives Firm Responses to Environmental Regulation?," in *Environmentally Responsible Supply Chains*, ed. Atalay Atasu (Springer, 2016), https://www.hbs.edu/faculty/Pages/item.aspx?num =49096.

2. "Changes in Enforcement by MA Department of Environmental Protection Over Time," AMEND: Archive of Massachusetts ENvironmental Data, last modified August 13, 2017, https://openamend.org /2017/04/02/dep-enforcements.html.

3. Johannes Reiche, "AI System Predicts Illegal Deforestation: Already Prevented the Clearing of 30 Hectares Near a Gold Mine," Wageningen University & Research, December 5, 2023, https://www.wur.nl /en/research-results/research-institutes/environmental-research/show -wenr/ai-system-predicts-illegal-deforestation-already-prevented-the -clearing-of-30-hectares-near-a-gold-mine.htm.

4. Siti Nadhirah Zainurin et al., "Advancements in Monitoring Water Quality Based on Various Sensing Methods: A Systematic Review," *International Journal of Environmental Research and Public Health* 19, no. 21 (2022): article 14080, https://doi.org/10.3390/ijerph192114080.

5. Joshua O. Ighalo, Adewale George Adeniyi, and Gonçalo Marques, "Artificial Intelligence for Surface Water Quality Monitoring and Assessment: A Systematic Literature Analysis," *Modeling Earth Systems and Environment* 7, no. 2 (2021): 669–81, https://doi.org/10.1007 /s40808-020-01041-z.

6. Ben Chugg et al., "Detecting Environmental Violations with Satellite Imagery in Near Real Time: Land Application Under the Clean Water Act," arXiv, 2022, https://doi.org/10.1145/3511808.3557104.

7. Jon Truby et al., "Banking on AI: Mandating a Proactive Approach to AI Regulation in the Financial Sector," *Law And Financial Markets Review* 14, no. 2 (2020): 110–20, https://doi.org/10.1080/17521440 .2020.1760454.

8. Cam Wilson, "AI Worse Than Humans in Every Way at Summarising Information, Government Trial Finds," *Crikey*, September 3, 2024, https://www.crikey.com.au/2024/09/03/ai-worse-summarising -information-humans-government-trial/.

9. "Compliance Management Software Market Size & Forecast," Verified Market Research, May 2024, https://www.verifiedmarketresearch .com/product/compliance-management-software-market/.

10. "Five Areas AI Could Transform Compliance and Risk Management," October 31, 2023, Moody's, https://www.moodys.com/web/en /us/kyc/resources/insights/five-areas-ai-could-transform-compliance -and-risk-management.html; Todd Ehret, "Where AI Will Play an

Important Role in Governance, Risk & Compliance Programs," Thomson Reuters, August 24, 2023, https://www.thomsonreuters.com/en-us /posts/corporates/ai-governance-risk-compliance-programs/; "Norm Ai," accessed March 26, 2025, https://www.norm.ai/.

11. "Whistleblower Program," US Securities and Exchange Commission, October 10, 2024, https://www.sec.gov/enforcement-litigation /whistleblower-program.

12. Alexander I. Platt, "The Whistleblower Industrial Complex," *Yale Journal on Regulation* 40, no. 2 (2023), https://www.yalejreg.com /print/the-whistleblower-industrial-complex/.

13. Matías Dewey and Donato Di Carlo, "Governing Through Non-enforcement: Regulatory Forbearance as Industrial Policy in Advanced Economies," *Regulation & Governance* 16, no. 3 (2021): 930–50, https:// doi.org/10.1111/rego.12382.

14. Andreas Fiebelkorn, "State Capture Analysis: How to Quantitatively Analyze the Regulatory Abuse by Business-State Relationships," Report, The International Bank for Reconstruction and Development/ The World Bank, 2019, https://documents1.worldbank.org/curated /en/785311576571172286/pdf/State-Capture-Analysis-How-to-Quan titatively-Analyze-the-Regulatory-Abuse-by-Business-State-Relation ships.pdf.

CHAPTER 26

1. Seth Lazar, "Legitimacy, Authority, and Democratic Duties of Explanation," in *Oxford Studies in Political Philosophy*, vol. 10, ed. David Sobel and Steven Wall (Oxford University Press, 2024), 28–56, https://doi.org/10.1093/oso/9780198909460.003.0002.

2. Cuong Tran et al., "Decision Making with Differential Privacy Under a Fairness Lens," in *Proceedings of the Thirtieth International Joint Conference on Artificial Intelligence*, ed. Zhi-Hua Zhou (International Joint Conferences on Artificial Intelligence, 2021), 560–66, https:// doi.org/10.24963/ijcai.2021/78.

CHAPTER 27

1. Dan Goodin, "Crooks Plant Backdoor in Software Used by Courtrooms Around the World," *Ars Technica*, May 23, 2024, https://

arstechnica.com/security/2024/05/crooks-plant-backdoor-in-software
-used-by-courtrooms-around-the-world/.

2. Bruce Schneier, "Every Part of the Supply Chain Can Be Attacked,"
New York Times, September 25, 2019, https://www.nytimes.com/2019
/09/25/opinion/huawei-internet-security.html.

3. "How America Built an AI Tool to Predict Taliban Attacks," *Econo-
mist*, July 31, 2024, https://www.economist.com/science-and-technol
ogy/2024/07/31/how-america-built-an-ai-tool-to-predict-taliban
-attacks.

4. Kevin Eykholt et al., "Robust Physical-World Attacks on Deep
Learning Models," arXiv, 2017, https://doi.org/10.48550/arxiv.1707
.08945.

5. Milad Nasr et al., "Scalable Extraction of Training Data from (Production) Language Models," arXiv, 2023, https://doi.org/10.48550
/arxiv.2311.17035.

6. Bruce Schneier and Davi Ottenheimer, "Web 3.0 Requires Data
Integrity," *Communications of the ACM*, March 24, 2025, https://doi
.org/10.1145/3723438.

CHAPTER 28

1. Sayash Kapoor et al., "Promises and Pitfalls of Artificial Intelligence for Legal Applications," *Journal of Cross-disciplinary Research in
Computational Law* 2, no. 2 (2024), https://journalcrcl.org/crcl/article
/view/62.

2. Lucian E. Dervan, "Fourteen Principles and a Path Forward for Plea
Bargaining Reform," *Criminal Justice Magazine*, January 22, 2024,
https://www.americanbar.org/groups/criminal_justice/resources
/magazine/2024-winter/fourteen-principles-path-forward-plea-bar
gaining-reform/.

3. Stephanos Bibas, "Incompetent Plea Bargaining and Extrajudicial
Reforms," *Harvard Law Review* 126, no. 1 (2012): 150–74, https://har
vardlawreview.org/print/vol-126/incompetent-plea-bargaining-and
-extrajudicial-reforms/.

4. Larry Neumeister, "An AI Avatar Tried to Argue a Case Before a
New York Court. The Judges Weren't Having It," AP News, April 4,

2025, https://apnews.com/article/artificial-intelligence-ai-courts-nyc
-5c97cba3f3757d9ab3c2e5840127f765.

5. Jim Ash, "Miami-Dade Public Defender Is Using Artificial Intelligence for Research and Case Preparation," *Florida Bar News*, December
26, 2023, https://www.floridabar.org/the-florida-bar-news/miami-dade
-public-defender-is-using-artificial-intelligence-for-research-and-for
-case-preparation/.

6. Jim Ash, "High Costs and High Turnover: Miami-Dade State
Attorney and PD Seek Solutions," *Florida Bar News*, March 22, 2024,
https://www.floridabar.org/the-florida-bar-news/high-costs-and-high
-turnover-miami-dade-state-attorney-and-pd-seek-solutions/.

7. Hannah L. Shotton, "Internet Frisking Jurors During Voir Dire: The
Case for Imposing Judicial Limitations," *Liberty University Law Review*
18, no. 3 (2024), Scholars Crossing, n.d., https://digitalcommons.liberty
.edu/lu_law_review/vol18/iss3/4.

8. Marcela Ayres and Bernardo Caram, "Brazil Hires OpenAI to Cut
Costs of Court Battles," Reuters, June 11, 2024, https://www.reuters
.com/technology/artificial-intelligence/brazil-hires-openai-cut-costs
-court-battles-2024-06-11/.

9. Cory Doctorow, "Commentary by Cory Doctorow: Plausible Sentence Generators," *Locus*, September 4, 2023, https://locusmag.com
/2023/09/commentary-by-cory-doctorow-plausible-sentence-generators/.

10. Yun-Chien Chang and Daniel Klerman, "Settlement Around the
World: Settlement Rates in the Largest Economies," *Journal of Legal
Analysis* 14, no. 1 (2022): 80–175, https://doi.org/10.1093/jla/laac006.

CHAPTER 29

1. Barnhart v. Thomas, 540 U.S. 20 (2003).

2. Kurt Glaze et al., "Artificial Intelligence for Adjudication," in *The
Oxford Handbook of AI Governance*, ed. Justin B. Bullock (Oxford University Press, 2022), 779–96, https://doi.org/10.1093/oxfordhb/9780
197579329.013.46.

3. Michael Broyde and Yiyang Mei, "Don't Kill the Baby: The Case for
AI in Arbitration," arXiv, 2024, https://doi.org/10.48550/arxiv.2408
.11608.

4. David A. Hoffman and Yonathan A. Arbel, "Generative Interpretation," *New York University Law Review* 99, no. 2 (2024): 451–514, https://doi.org/10.2139/ssrn.4526219.

5. Brady Williams, "Unconscionability as a Sword: The Case for an Affirmative Cause of Action," *California Law Review* 107, no. 6 (2019): 2015–70, https://doi.org/10.15779/z382b8vc3w.

6. Catherine A. Rogers, "The Politics of International Investment Arbitrators," *Santa Clara International Law Review* 12 (2013): 217–56, https://ssrn.com/abstract=2347843.

7. Catherine A. Rogers, "Reconceptualizing the Party-Appointed Arbitrator and the Meaning of Impartiality," *Harvard International Law Journal* 64, no. 1 (2023): 137–202, https://journals.law.harvard.edu/ilj/2023/09/reconceptualizing-the-party-appointed-arbitrator-and-the-meaning-of-impartiality/.

CHAPTER 30

1. Paul Ohm, "The Computer Fraud and Abuse Act After Van Buren," *American Constitution Society Supreme Court Review* 5 (2021), https://www.acslaw.org/analysis/acs-journal/2020-2021-acs-supreme-court-review/the-computer-fraud-and-abuse-act-after-van-buren/.

2. Lisa Eckelbecker and Bob Kievra, "Sandwich Skirmish: White City Eatery Loses Bid to Block Rival's Opening," *Worcester Telegram & Gazette*, November 10, 2006, https://www.telegram.com/story/news/local/east-valley/2006/11/11/sandwich-skirmish/53019621007/; Marjorie Florestal, "Is a Burrito a Sandwich? Exploring Race, Class, and Culture in Contracts," *Michigan Journal of Race and Law* 14, no. 1 (2008), https://repository.law.umich.edu/mjrl/vol14/iss1/1/.

3. Anna Price, "As a Matter of Law, Is a Taco a Sandwich?," *In Custodia Legis* (blog), May 21, 2024, https://blogs.loc.gov/law/2024/05/as-a-matter-of-law-is-a-taco-a-sandwich/; Ilya Somin, "Indiana Court Rules Burritos and Tacos Qualify as Sandwiches," *The Volokh Conspiracy* (blog), May 19, 2024, https://reason.com/volokh/2024/05/19/indiana-court-rules-burritos-and-tacos-qualify-as-sandwiches/.

4. David Cole, "Obamacare Upheld: How and Why Did Justice Roberts Do It?," *Nation*, June 28, 2012, https://www.thenation.com/article/archive/obamacare-upheld-how-and-why-did-justice-roberts-do-it/.

5. Jack Kieffaber, "Predictability, AI, and Judicial Futurism: Why Robots Will Run the Law and Textualists Will Like It," SSRN, November 6, 2024, https://doi.org/10.2139/ssrn.4966334; Christoph Engel and Richard H. McAdams, "Asking GPT for the Ordinary Meaning of Statutory Terms," MPI Collective Goods Discussion Paper no. 2024/5, February 6, 2024, https://doi.org/10.2139/ssrn.4718347.

6. Nate Raymond, "US Judge Makes 'Unthinkable' Pitch to Use AI to Interpret Legal Texts," Reuters, May 29, 2024, https://www.reuters.com /legal/transactional/us-judge-makes-unthinkable-pitch-use-ai-inter pret-legal-texts-2024-05-29/.

7. Primavera De Filippi and Samer Hassan, "Blockchain Technology as a Regulatory Technology: From Code Is Law to Law Is Code," *First Monday* 21, no. 12 (2016), https://doi.org/10.5210/fm.v21i12.7113.

CHAPTER 31

1. Luke Taylor, "Colombian Judge Says He Used ChatGPT in Ruling," *Guardian*, February 2, 2023, https://www.theguardian.com/technology /2023/feb/03/colombia-judge-chatgpt-ruling; James Titcomb, "Judges Given Green Light to Use ChatGPT in Legal Rulings," *Telegraph*, December 11, 2023, https://www.telegraph.co.uk/business/2023/12/12 /judges-given-green-light-use-chatgpt-legal-rulings/.

2. Todd C. Peppers and Christopher Zorn, "Law Clerk Influence on Supreme Court Decision Making: An Empirical Assessment," *DePaul Law Review* 58, no. 1 (2008), https://via.library.depaul.edu/law-review /vol58/iss1/3/; Sara C. Benesh, David A. Armstrong, and Zachary Wallander, "Advisors to Elites," *Journal of Law and Courts* 8, no. 1 (2020): 51–73, https://doi.org/10.1086/704740.

3. William H. Rehnquist, "Who Writes Decisions of the Supreme Court," *Brief* 53 (1957–1958), https://heinonline.org/HOL/Landing Page?handle=hein.journals/briephid53&div=10&id=&page=.

4. Paul J. Wahlbeck, James F. Spriggs II, and Lee Sigelman. "Ghostwriters on the Court?" *American Politics Research*, 2002, https://doi.org /10.1177/1532673x02030002003.

5. Corey Ditslear and Lawrence Baum, "Selection of Law Clerks and Polarization in the U.S. Supreme Court," *Journal of Politics* 63, no. 3 (2001): 869–85, https://doi.org/10.1111/0022-3816.00091.

6. Melissa Heikkilä, "AI Language Models Are Rife with Different Political Biases," *MIT Technology Review*, August 7, 2023, https://www .technologyreview.com/2023/08/07/1077324/ai-language-models-are -rife-with-political-biases/.

7. Will Knight, "Meet ChatGPT's Right-Wing Alter Ego," *Wired*, April 27, 2023, https://www.wired.com/story/fast-forward-meet-chatgpts -right-wing-alter-ego/.

8. Josef Valvoda et al., "The Ethics of Automating Legal Actors," *Transactions of the Association for Computational Linguistics* 12 (2024): 700–20, https://doi.org/10.1162/tacl_a_00668.

9. Neel Guha et al., "LegalBench: A Collaboratively Built Benchmark for Measuring Legal Reasoning in Large Language Models," arXiv, 2023, https://arxiv.org/abs/2308.11462.

10. "For the First Time, a Dutch Judge Uses ChatGPT in a Verdict, Experts Are Shocked," *NL Times*, August 3, 2024, https://nltimes.nl /2024/08/03/first-time-dutch-judge-uses-chatgpt-verdict-experts -shocked.

CHAPTER 32

1. Bruce Schneier, "The Coming AI Hackers," The Belfer Center for Science and International Affairs, April 2021, https://www.belfercenter .org/publication/coming-ai-hackers.

2. Jowi Morales, "AI Models That Cost $1 Billion to Train Are Underway, $100 Billion Models Coming—Largest Current Models Take 'Only' $100 Million to Train," *Tom's Hardware*, July 7, 2024, https:// www.tomshardware.com/tech-industry/artificial-intelligence/ai -models-that-cost-dollar1-billion-to-train-are-in-development-dollar 100-billion-models-coming-soon-largest-current-models-take-only -dollar100-million-to-train-anthropic-ceo.

3. "CIP and Anthropic Launch Collective Constitutional AI," *The Collective Intelligence Project* (blog), accessed March 29, 2025, https:// www.cip.org/blog/ccai.

4. Wojciech Zaremba et al., "Democratic Inputs to AI," OpenAI, May 25, 2023, https://openai.com/index/democratic-inputs-to-ai/.

5. Roel Nahuis and Harro Van Lente, "Where Are the Politics? Perspectives on Democracy and Technology," *Science Technology &*

Human Values 33, no. 5 (2008): 559–81, https://doi.org/10.1177/016 2243907306700.

CHAPTER 33

1. Bruce Schneier, *Liars and Outliers: Enabling the Trust That Society Needs to Thrive* (John Wiley & Sons, 2012).

2. Bruce Schneier, "AI and Trust," The Belfer Center for Science and International Affairs, November 27, 2023, https://www.belfercenter .org/publication/ai-and-trust.

3. Judith Donath and Bruce Schneier, "It's the End of the Web as We Know It," *Atlantic*, April 22, 2024, https://www.theatlantic.com /technology/archive/2024/04/generative-ai-search-llmo/678154/.

CHAPTER 34

1. Ben Cohen, "There's a New Hit Podcast That Will Blow Your Mind," *Wall Street Journal*, October 4, 2024, https://www.wsj.com /tech/ai/google-notebooklm-ai-podcast-deep-dive-audio-c30a06b3.

2. Jennifer Smith, "In Melrose, an Experiment in Hyper-local AI Podcasting," *CommonWealth Beacon*, October 17, 2024, https://common wealthbeacon.org/media/in-melrose-an-experiment-in-hyper-local -ai-podcasting/.

3. Reece Rogers, "Google's NotebookLM Now Lets You Customize Its AI Podcasts," *Wired*, October 17, 2024, https://www.wired.com/story /google-notebooklm-customize-ai-podcast/.

4. Stuart A. Thompson, "AI-Generated Content Discovered on News Sites, Content Farms and Product Reviews," *New York Times*, May 19, 2023, https://www.nytimes.com/2023/05/19/technology/ai-generated -content-discovered-on-news-sites-content-farms-and-product-reviews .html.

5. Ivan Mehta, "Artifact News App Now Uses AI to Rewrite Headline of a Clickbait Article," *TechCrunch*, June 2, 2023, https://techcrunch .com/2023/06/02/artifact-news-app-now-uses-ai-to-rewrite-headline -of-a-clickbait-article/.

6. Ross Williams, "Georgia Political Campaigns Start to Deploy AI but Humans Still Needed to Press the Flesh," *News from the States*,

April 25, 2024, https://www.newsfromthestates.com/article/georgia-political-campaigns-start-deploy-ai-humans-still-needed-press-flesh.

7. WashPostPR, "The Washington Post Launches a New Commenting Experience Exclusively for Subscribers," *Washington Post*, December 16, 2024, https://www.washingtonpost.com/pr/2024/12/16/washington-post-launches-new-commenting-experience-exclusively-subscribers/.

8. Damien S. Pfister, "The Logos of the Blogosphere: Flooding the Zone, Invention, and Attention in the Lott Imbroglio," *Argumentation and Advocacy* 47, no. 3 (2011): 141–62, https://digitalcommons.unl.edu/cgi/viewcontent.cgi?article=1004&context=commstudiespapers.

CHAPTER 35

1. Dermot Murphy, "When Local Papers Close, Costs Rise for Local Governments," *Columbia Journalism Review*, June 27, 2018, https://www.cjr.org/united_states_project/public-finance-local-news.php.

2. "Methodology," Global Right to Information Rating, accessed March 29, 2025, https://www.rti-rating.org/methodology/.

3. Sarah Ulrich, "Using Data to Expose Systemic Gender-Based Violence—in 10 Steps," Global Investigative Journalism Network, February 3, 2025, https://gijn.org/stories/10-steps-data-expose-gender-violence-eswatini/.

4. Vittoria Elliott, "Election Workers Are Drowning in Records Requests. AI Chatbots Could Make It Worse," *Wired*, April 10, 2024, https://www.wired.com/story/ai-chatbots-foia-requests-election-workers/.

5. Soubhik Barari and Tyler Simko, "LocalView, a Database of Public Meetings for the Study of Local Politics and Policy-making in the United States," *Scientific Data* 10, no. 1 (2023), https://doi.org/10.1038/s41597-023-02044-y.

6. "Data Sources & Methodology," Digital Democracy, accessed March 29, 2025, https://digitaldemocracy.calmatters.org/data-sources-methodology.

7. "What Is the Polarization Research Lab?," Polarization Research Lab, accessed March 29, 2025, https://americaspoliticalpulse.com/about/.

8. Hannah Natanson et al., "Elon Musk's DOGE Is Feeding Sensitive Federal Data into AI to Target Cuts," *Washington Post*, February 6,

2025, https://www.washingtonpost.com/nation/2025/02/06/elon-musk
-doge-ai-department-education/.

9. Paulo Savaget, Tulio Chiarini, and Steve Evans, "Empowering Political Participation Through Artificial Intelligence," *Science and Public Policy* 46, no. 3 (2019): 369–80, https://doi.org/10.1093/scipol/scy064.

10. Fernanda Odilla and Alice Mattoni, "Unveiling the Layers of Data Activism: The Organising of Civic Innovation to Fight Corruption in Brazil," *Big Data & Society* 10, no. 2 (2023), https://doi.org/10.1177/20539517231190078.

11. Paroma Soni, "A New Tool Allows Journalists to Quickly Sort Through FOIA Data Dumps," *Columbia Journalism Review*, March 2, 2022, https://www.cjr.org/innovations/gumshoe-foia-data-dumps-ai.php.

12. "AI and What Is True," Human Rights Data Analysis Group, 2024, https://hrdag.org/report/2024-review/.

13. Tom Phillips and Patricia Torres, "'Being on Camera Is No Longer Sensible': Persecuted Venezuelan Journalists Turn to AI," *Guardian*, August 27, 2024, https://www.theguardian.com/world/article/2024/aug/27/venezuela-journalists-nicolas-maduro-artificial-intelligence-media-election.

14. Nathan E. Sanders, "Legislative Comment Summarization Using AI Agents," GitHub, accessed March 29, 2025, https://github.com/nesanders/ai_comment_summarization_hria.

CHAPTER 36

1. Richard Wike and Janell Fetterolf, "Global Public Opinion in an Era of Democratic Anxiety," Pew Research Center, December 17, 2021, https://www.pewresearch.org/global/2021/12/07/global-public-opinion-in-an-era-of-democratic-anxiety/.

2. Michael Henry Tessler et al., "AI Can Help Humans Find Common Ground in Democratic Deliberation," *Science* 386, no. 6719 (2024), https://doi.org/10.1126/science.adq2852.

3. Spandana Singh, "Everything in Moderation," New America, last updated July 22, 2019, https://www.newamerica.org/oti/reports/everything-moderation-analysis-how-internet-platforms-are-using-artificial-intelligence-moderate-user-generated-content/.

4. Chris Horton, "The Simple but Ingenious System Taiwan Uses to Crowdsource Its Laws," *MIT Technology Review*, August 21, 2018, https://www.technologyreview.com/2018/08/21/240284/the-simple -but-ingenious-system-taiwan-uses-to-crowdsource-its-laws/.

5. James O'Donnell, "A Small US City Experiments with AI to Find Out What Residents Want," *MIT Technology Review*, April 15, 2025, https://www.technologyreview.com/2025/04/15/1115125/a-small-us -city-experiments-with-ai-to-find-out-what-residents-want.

6. Zoey Tseng, "Amplifying Voices: Talk to the City in Taiwan," AI Objectives Institute, last updated May 30, 2024, https://ai.objectives .institute/blog/amplifying-voices-talk-to-the-city-in-taiwan/.

7. Louis Rosenberg, "Artificial Swarm Intelligence vs. Human Experts," in *2016 International Joint Conference on Neural Networks (IJCNN)* (Curran Associates, 2017), 2547–51, https://doi.org/10.1109/ijcnn.2016 .7727517.

8. "Hearing the Humanity: Cortico's Collaboration with NPR," Cortico, accessed March 29, 2025, https://cortico.ai/news/hearing-the -humanity-cortico-s-collaboration-with-npr/.

9. Laura Giesen, "One Third of Eligible Youth Participates in Helsinki Youth Budget," *Democracy Technologies*, April 27, 2023, https://demo cracy-technologies.org/participation/one-third-of-eligible-youth -participates-in-helsinki-youth-budget/.

10. Rick Harrison, "Can Citizens' Assemblies Help Restore Trust in Government?," Yale University, ISPS (blog), November 30, 2023, https://isps.yale.edu/news/blog/2023/11/can-citizens-assemblies-help -restore-trust-in-government.

11. Didier Caluwaerts et al., "Deliberation and Polarization: A Multi-disciplinary Review," *Frontiers in Political Science* 5 (2023), https://doi .org/10.3389/fpos.2023.1127372.

12. Diego Garzia et al., "Affective Polarization in Comparative and Longitudinal Perspective," *Public Opinion Quarterly* 87, no. 1 (2023): 219–31, https://doi.org/10.1093/poq/nfad004.

13. Dan Jones, "Walmart Workers Are Using AI in Their Fight for Respect," Kairos Center, November 21, 2016, https://kairoscenter.org /walmart-workers-artificial-intelligence/; Frances Flanagan and Michael Walker, "How Can Unions Use Artificial Intelligence to Build Power?

The Use of AI Chatbots for Labor Organizing in the US and Australia," *New Technology, Work and Employment* 36, no. 2 (2021): 159–76, https://doi.org/10.1111/ntwe.1217.

14. "Unions Mobilise AI to Turn the Tables on Wage Theft in Hospitality," United Workers Union, accessed March 29, 2025, https://unitedworkers.org.au/media-release/unions-mobilise-ai-to-turn-the-tables-on-wage-theft-in-hospitality/.

15. "Technology for the Labor Movement," The Workers Lab, accessed March 29, 2025, https://www.theworkerslab.com/technology.

16. Bradford J. Kelley, "Belaboring the Algorithm: Artificial Intelligence and Labor Unions," *Yale Journal on Regulation*, June 15, 2024, https://www.yalejreg.com/bulletin/belaboring-the-algorithm-artificial-intelligence-and-labor-unions/.

17. Brian Merchant, "Learning from the Luddites," *Boston Review*, December 4, 2024, https://www.bostonreview.net/forum_response/learning-from-the-luddites/.

CHAPTER 37

1. Melanie Dione, "Resistbot at Six: Building a Community," Resistbot, March 8, 2023, https://resist.bot/news/2023/03/08/resistbot-at-six-building-a-community.

2. Jason Putorti, "Introducing the AI Writer," Resistbot, July 4, 2023, https://resist.bot/news/2023/07/04/introducing-the-ai-writer.

3. Weixin Liang et al., "The Widespread Adoption of Large Language Model-Assisted Writing Across Society," arXiv, 2025, https://doi.org/10.48550/arxiv.2502.09747.

4. Steven Overly, "How AI Is Transforming a Lawmaker's Life After a Terrible Diagnosis," *Politico*, September 9, 2024, https://www.politico.com/news/magazine/2024/09/09/ai-jennifer-wexton-qa-00177949.

5. Chad Orzel, "Consensus by Attrition," *Counting Atoms* (blog), June 26, 2023, https://chadorzel.substack.com/p/consensus-by-attrition.

6. Alicia Combaz et al., "Applications of Artificial Intelligence Tools to Enhance Legislative Engagement: Case Studies from Make.Org and MAPLE," arXiv, February 12, 2025, https://arxiv.org/abs/2503.04769.

7. Johanna Habicht et al., "Closing the Accessibility Gap to Mental Health Treatment with a Personalized Self-Referral Chatbot," *Nature Medicine* 30, no. 2 (2024): 595–602, https://doi.org/10.1038/s41591-023-02766-x.

8. "Sign Up for SNAP (Formerly Known as Food Stamps)," mRelief, accessed March 29, 2025, https://mrelief.com/.

9. "iMMPATH," Justicia Lab, accessed March 29, 2025, https://immpath.ai/.

10. Pamela Herd and Donald P. Moynihan, *Administrative Burden: Policymaking by Other Means* (Russell Sage Foundation, 2019).

CHAPTER 38

1. Chen Wang, "Outsourcing Voting to AI: Can ChatGPT Advise Index Funds on Proxy Voting Decisions?," *Fordham Journal of Corporate & Financial Law* 29, no. 1 (2023), 113–89, https://ir.lawnet.fordham.edu/jcfl/vol29/iss1/3/.

2. Michiel Bakker et al., "Fine-Tuning Language Models to Find Agreement Among Humans with Diverse Preference," in *Advances in Neural Information Processing Systems 35 (NeurIPS 2022)*, ed. Sanmi Koyejo et al. (Curran Associates, 2023), 38176–38189, https://proceedings.neurips.cc/paper/2022/hash/f978c8f3b5f399cae464e85f72e28503-Abstract-Conference.html; Sara Fish et al., "Generative Social Choice," arXiv, 2023, https://doi.org/10.48550/arxiv.2309.01291.

3. Konstantin Scheuermann and Angela Aristidou, "Could AI Speak on Behalf of Future Humans?," *Stanford Social Innovation Review*, February 5, 2024, https://ssir.org/articles/entry/ai-voice-collective-decision-making.

4. Charlotte Mol, "Children's Representation in Family Law Proceedings," *International Journal of Children's Rights* 27, no. 1 (2019): 66–98, https://doi.org/10.1163/15718182-02701001.

5. "Dashboard," Eco Jurisprudence Monitor, accessed March 30, 2025, https://ecojurisprudence.org/dashboard/?map-style=political.

6. Arik Kershenbaum, "The Race to Translate Animal Sounds into Human Language," *Wired*, December 22, 2024, https://www.wired.com/story/artificial-intelligence-translation-animal-sounds-human-language/.

CHAPTER 39

1. Charles Duhigg, "Silicon Valley, the New Lobbying Monster," *New Yorker*, October 7, 2024, https://www.newyorker.com/magazine/2024/10/14/silicon-valley-the-new-lobbying-monster.

CHAPTER 40

1. "We All Can't Support This: 3rd Draft of the EU AI Act's GPAI Code of Practice. Joint Statement," European Writers' Council, March 28, 2025, https://europeanwriterscouncil.eu/2503_nosupport_cop/.

2. "EU's AI Act Fails to Set Gold Standard for Human Rights," European Digital Rights (EDRi), April 3, 2024, https://edri.org/our-work/eu-ai-act-fails-to-set-gold-standard-for-human-rights/.

3. Eleni Courea, "UK Delays Plans to Regulate AI as Ministers Seek to Align with Trump Administration," *Guardian*, February 24, 2025, https://www.theguardian.com/technology/2025/feb/24/uk-delays-plans-to-regulate-ai-as-ministers-seek-to-align-with-trump-administration.

4. Pieter Haeck, "EU Rules for Advanced AI Are Step in Wrong Direction, Google Says," *Politico*, February 10, 2025, https://www.politico.eu/article/google-eu-rules-advanced-ai-artificial-intelligence-step-in-wrong-direction/.

5. Sigal Samuel et al., "California's Governor Has Vetoed a Historic AI Safety Bill," *Vox*, September 29, 2024, https://www.vox.com/future-perfect/369628/ai-safety-bill-sb-1047-gavin-newsom-california.

6. "SEA-LION.AI—South East Asian Languages in One Network," AI Singapore, accessed March 30, 2025, https://sea-lion.ai/.

7. "Indonesia's Indosat, GoTo Launch Local-language AI Model," Reuters, November 13, 2024, https://www.reuters.com/technology/artificial-intelligence/indonesias-indosat-goto-launch-local-language-ai-model-2024-11-14/.

8. Jennifer Wang and Mark Muro, "How the National Artificial Intelligence Research Resource Can Pilot Inclusive AI," Brookings Institute, July 9, 2024, https://www.brookings.edu/articles/how-the-national-artificial-intelligence-research-resource-can-pilot-inclusive-ai/.

9. "French Government Will Use AI to Modernise Public Services," *RFI*, April 23, 2024, https://www.rfi.fr/en/france/20240423-french -government-will-use-ai-to-modernise-public-services.

10. Argonne National Laboratory, "Argonne and Riken Sign a Memorandum of Understanding in Support of AI for Science," news release, April 12, 2024.

11. Zeyi Yang, "Here's How DeepSeek Censorship Actually Works— and How to Get Around It," *Wired*, January 31, 2025, https://www .wired.com/story/deepseek-censorship/.

12. Alex Pascal and Nathan E. Sanders, "Why US States Are the Best Labs for Public AI," *Tech Policy Press*, April 3, 2025, https://www.tech policy.press/why-us-states-are-the-best-labs-for-public-ai/.

13. Kelsey Piper, "Inside OpenAI's Multibillion-Dollar Gambit to Become a For-Profit Company," *Vox*, October 28, 2024, https://www .vox.com/future-perfect/380117/openai-microsoft-sam-altman-non profit-for-profit-foundation-artificial-intelligence.

14. Dylan Patel, "Google 'We Have No Moat, and Neither Does OpenAI,'" SemiAnalysis, May 4, 2023, https://www.semianalysis.com /p/google-we-have-no-moat-and-neither.

15. Michelle Toh, "Jack Ma Loses More Than Half of His Wealth After Criticizing Chinese Regulators," CNN, July 12, 2023.

16. Masha Gessen, "The Wrath of Putin," *Vanity Fair*, April 2012, https://archive.vanityfair.com/article/share/a35d8805-2ceb-4c03 -9ece-595e23d0cf52.

17. "Russia Puts Exiled Tycoon and Opposition Leader Khodorkovsky on Wanted List for War Comments," AP News, January 9, 2024, https://apnews.com/article/mikhail-khodorkovsky-russia-putin-b59 a782cf1802c0f37c7bb9a0c98139a.

CHAPTER 41

1. Becca Ricks et al., "Creating Trustworthy AI," Mozilla, December 2020, https://foundation.mozilla.org/en/insights/trustworthy-ai-white paper/.

2. Michele Gilman, "Democratizing AI: Principles for Meaningful Public Participation," Data & Society, September 27, 2023, https://

datasociety.net/library/democratizing-ai-principles-for-meaningful
-public-participation/.

3. "Going Public: Exploring Public Participation in Commercial AI Labs," Ada Lovelace Institute, December 2023, https://www.adalove laceinstitute.org/report/going-public-participation-ai/.

4. "A Roadmap to Democratic AI," The Collective Intelligence Project, March 2024, https://www.cip.org/research/ai-roadmap.

5. "Trustworthy and Responsible AI," NIST, accessed March 30, 2025, https://www.nist.gov/trustworthy-and-responsible-ai.

6. "The Foundation Model Transparency Index," Center for Research on Foundation Models, May 2024, https://crfm.stanford.edu/fmti /May-2024/index.html.

7. Karen Hao, "OpenAI Is Giving Microsoft Exclusive Access to Its GPT-3 Language Model," *MIT Technology Review*, September 23, 2020, https://www.technologyreview.com/2020/09/23/1008729/openai-is -giving-microsoft-exclusive-access-to-its-gpt-3-language-model/; Charles Duhigg, "The Inside Story of Microsoft's Partnership with OpenAI," *New Yorker*, December 1, 2023, https://www.newyorker. com/magazine/2023/12/11/the-inside-story-of-microsofts-partner ship-with-openai.

8. Edward Zitron, "How Does OpenAI Survive?," *Where's Your Ed At* (blog), July 29, 2024, https://www.wheresyoured.at/to-serve-altman/.

9. Stephanie Palazzolo and Cory Weinberg, "OpenAI Plots Charging $20,000 a Month for PhD-Level Agents," *Information*, March 5, 2025, https://www.theinformation.com/articles/openai-plots-charging-20 -000-a-month-for-phd-level-agents.

10. Zvi Mowshowitz, "How AI Chatbots Became Political," *New York Times*, March 28, 2024, https://www.nytimes.com/interactive/2024 /03/28/opinion/ai-political-bias.html.

11. "Meta: Llama 2," Center for Research on Foundation Models, May 2024, https://crfm.stanford.edu/fmti/May-2024/company-reports/Meta _Llama%202.html.

CHAPTER 42

1. Bruce Schneier and Nathan Sanders, "The AI Wars Have Three Factions, and They All Crave Power," *New York Times*, September 28,

2023, https://www.nytimes.com/2023/09/28/opinion/ai-safety-ethics
-effective.html.

2. Danielle Allen, "Is Fixing Democracy Partisan? Here Are Answers
to This and More Questions," *Washington Post*, August 23, 2023,
https://www.washingtonpost.com/opinions/2023/08/23/democracy
-renovation-faq-danielle-allen/.

3. Molly Kinder, "Hollywood Writers Went on Strike to Protect Their
Livelihoods from Generative AI. Their Remarkable Victory Matters
for All Workers," Brookings, April 12, 2024, https://www.brookings
.edu/articles/hollywood-writers-went-on-strike-to-protect-their-live
lihoods-from-generative-ai-their-remarkable-victory-matters-for-all
-workers/.

4. "Decoding the White House AI Executive Order's Achievements,"
Stanford HAI, November 2, 2023, https://hai.stanford.edu/news/decod
ing-white-house-ai-executive-orders-achievements.

5. Ege Erdil, "What Went into Training DeepSeek-R1?," Epoch AI,
January 31, 2025, https://epoch.ai/gradient-updates/what-went-into
-training-deepseek-r1.

6. "Approval of the Content of the Draft Communication from the
Commission: Commission Guidelines on Prohibited Artificial Intelligence Practices Established by Regulation (Eu) 2024/1689 (AI Act),"
European Commission, February 4, 2025, https://eimin.lrv.lt/public
/canonical/1738843867/5382/Gairiu%20projektas.pdf.

7. Bruce Schneier and Nathan Sanders, "We Don't Need to Reinvent
Our Democracy to Save It from AI," The Belfer Center for Science and
International Affairs, February 9, 2023, https://www.belfercenter.org
/publication/we-dont-need-reinvent-our-democracy-save-it-ai.

INDEX